# Greenbelt, Maryland

Creating the North American Landscape

*Gregory Conniff, Edward K. Muller, David Schuyler*
CONSULTING EDITORS

*George F. Thompson*
SERIES FOUNDER AND DIRECTOR

*Published in Cooperation with the Center for American Places,
Santa Fe, New Mexico, and Harrisonburg, Virginia*

Shopping center, movie theater, and apartments,
Greenbelt, 1942. Photo by Marjory Collins, U.S. Office
of War Information, Prints and Photographs Division,
Library of Congress

# Greenbelt, Maryland

A Living Legacy of the New Deal

Cathy D. Knepper

The Johns Hopkins University Press

BALTIMORE AND LONDON

© 2001 The Johns Hopkins University Press
All rights reserved. Published 2001
Printed in the United States of America on acid-free paper
9  8  7  6  5  4  3  2  1

The Johns Hopkins University Press
2715 North Charles Street
Baltimore, Maryland 21218-4363
www.press.jhu.edu

*Library of Congress Cataloging-in-Publication Data*
Knepper, Cathy D.
    Greenbelt, Maryland : a living legacy of the New Deal / Cathy D.
    Knepper.
    p.    cm. — (Creating the North American landscape)
    "Published in cooperation with the Center for American Places,
    Harrisonburg, Virginia, and Santa Fe, New Mexico."
    Includes bibliographical references and index.
    ISBN 0-8018-6490-9 (alk. paper)
1. Greenbelt (Md.)—History. 2. Planned communities—United States—
Case studies. I. Center for American Places. II. Title. III. Series.
F189.G7 K58 2001
975.2′51—dc21          00-008850

A catalog record for this book is available from the British Library.

To the people of Greenbelt,
past, present, and future

# Contents

# Acknowledgments

I WOULD LIKE TO THANK all the current and former Greenbelters who kindly let me inquire into their lives. Three individuals of the Greenbelt community have been of particular help to me: Betty Allen and Tom Simon, Tugwell Room librarians, and Dorothy Lauber, in the city manager's office. The assistance provided by the Beeke-Levy Research Fellowship allowed for my work at the Roosevelt Library in Hyde Park, New York. I owe a debt to Hasia Diner, who offered much-needed encouragement. Thanks go to Ann Denkler and Katie Scott-Childress, curators of the Greenbelt Museum, and Mary Lou Williamson, editor of the *Greenbelt News Review,* for their assistance with photographs. David Schuyler provided excellent editing advice. My husband and sons have exhibited great tolerance while learning more about Greenbelt than they could ever have wished, and thus deserve my gratitude.

# Introduction

THE CITY OF GREENBELT would appear to have a split personality: at its creation it was the most modern of places in terms of its architecture and planning, but at the same time it was the most traditional of places, a throwback to ideas of community in existence since America's earliest colonial days.

The fact that Greenbelt is a planned community is not, by itself, a mark of distinction. Planned communities have dotted the American landscape throughout the history of its settlement. Because of the cultural emphasis in the United States on individualism, acceptance of planned communities has languished in recent years. Further, the history of planned communities is little understood. As pointed out in 1974 by John Reps, "Only the arrogance of a people ignorant of their own tradition could create the prevailing, and quite mistaken, belief that planning new towns as an instrument of public policy is an invention of our own era."[1] Even though Reps limits himself to a discussion of new towns resulting from "public initiative," he gives dozens of examples dating to colonial times; most of these were in New England; others were Annapolis, Maryland, and Williamsburg, Virginia. Of course, Washington, D.C., remains a prominent example of a planned community, as do state capitals such as Columbia, Tallahassee, Raleigh, Indianapolis, Columbus, and Austin. "History is not usually thought of as a weapon," Reps says, "but it can be used as such, at least in the counterattack against the charge that publicly initiated new towns are somehow un-American and foreign to our tradition."[2]

Other than Washington, D.C., most governmental efforts in town planning have been carried out at the state level. The federal greenbelt town program of the 1930s, which created Greenbelt, Maryland, Greenhills, Ohio, and Greendale, Wisconsin, remains a prominent exception. Due to their unique place in planning history, the greenbelt towns have attracted attention from planners and historians. Recent scholarly works include *Main*

*Street Ready-Made,* by Arnold Alanen and Joseph Eden, about Greendale; Zane Miller, *Suburb: Neighborhood and Community in Forest Park, Ohio, 1935–1976,* regarding the planned community created from land originally allotted to Greenhills; and Susan Klaus compiled *Links in the Chain: Greenbelt, Maryland, and the New Town Movement in America.* Earlier works on the greenbelt program are Joseph Arnold's *The New Deal in the Suburbs* and Albert Mayer's *Greenbelt Towns Revisited.*[3]

My work, while fitting into the existing literature on greenbelt towns, is unique in its focus on the ideology of the greenbelt town program. This study evaluates the effect of time on the original goals and ideals of Greenbelt as articulated in 1937, looking at its unique physical design but focusing on the concept of cooperation—and thus both on the initial ideology presented by the federal government to the first residents and on their subsequent use of it.

To understand Greenbelt and its residents, two terms need to be defined: *planned community and cooperation.* For the purpose of this study, a planned community is one thought out and constructed as a conceptual whole, with the goal of providing an especially agreeable physical environment for its residents. According to Merriam-Webster's, *cooperation* is an "association of persons for common benefit," a definition that applies to Greenbelt. Furthermore, federal government officials initiated Greenbelt residents into the realm of *economic* cooperation by introducing cooperative stores into the community. They also introduced the concept of *social* cooperation, the idea that residents would work together to develop community organizations to meet their needs.

Greenbelters viewed cooperation broadly, as a way to achieve the long-term continuation of their communal goals. They did not see it as meaning everyone had to agree with everyone else at all times. Greenbelters frequently differed in their specific, short-term ideas of what would suit their town, but this was accepted and even welcomed as evidence of interest. Ultimately, to Greenbelters cooperation meant a devotion to organizing and running their own town.

From the beginning, people in Greenbelt felt they were taking part in a significant social experiment and thus kept many records of their activities. These, along with copies of the town's weekly newspaper and commentary on the city from Washington and Baltimore newspapers that have been collected and placed in the Tugwell Room in the Greenbelt branch of the Prince George's County Library, document Greenbelt's history. Contents of the Tugwell Room, the papers of those involved in the formation of Greenbelt, such as Rexford Tugwell and Jack Lansill, along with observations by numerous individuals, provide detailed information on the founding and subsequent development of the town. In addition, oral histories in written, cassette tape, and video form, many of which were produced as part of the town's fiftieth anniversary celebration, are also available. I conducted a number of interviews, attended town meetings and the Labor Day festival, and otherwise made an effort to become familiar with life in Greenbelt.

From the sources used, it can be seen that I have combined the methodologies of history and ethnography in my research. This combination is unusual, and I found it rather difficult to carry out, as conflicts occur between the two. History demands a clear, objective analysis of facts, whereas ethnography focuses on identification with the thoughts, feelings, and point of view of those being studied. I attempted to use each method in the most effective manner to benefit this study. My analysis of Greenbelt's plan and how it fared over time was, I hope, done in an objective—that is, historical—manner. On the other hand, I relied on the viewpoints of the people themselves in deciding what matters were vital to discuss. Instead of imposing an artificial framework on Greenbelters, I used their own: whatever was important to them was important to include. As much as possible, I let the people of Greenbelt tell their own story. They are articulate and eager to talk of life in their town. To them their history, so carefully preserved, serves as a guide for the present. Greenbelters eagerly express the desirability of life in a cooperative, planned community as they keep a remnant of the New Deal alive.

When Greenbelt was only one year old, Lewis Mumford in his 1938 work *The Culture of Cities* showcased it as an American example of proper planning. In the text accompanying an airplane view of the city under construction he summed up his views: "Much more compact than the scatter-building of the nineteenth century suburb; much more open than the traditional types of city design. Shows the great benefits obtainable only through comprehensive design, large scale planning, scientific appraisal of needs, and unified land-ownership and large-scale building operations. Communities of this order were first projected by Robert Owen: they have now become a universal indication of biotechnic city design."[4]

The showing of the movie *The City*, which was based on Mumford's book, at the 1939 New York City World's Fair gave Mumford's ideas wide exposure. Mumford himself wrote the commentary for the movie, which presented his ideas in a visually dramatic fashion. *The City* begins with a view of a quaint New England village, demonstrating how suited the built environment is for its residents. The commentator describes it this way: "The town was us and we were part of it. We never let our towns grow too big for us to manage." Accompanied by dramatic music composed by Aaron Copland, the scene switches to a man dwarfed by an enormous blast furnace. This insignificant-looking individual works next to a huge cauldron of liquid steel; at the end of the day, he trudges to his nearby home, black smoke from hundreds of smokestacks billowing in the background. The narrator points out: "There's prisons for a guy sent up for crime with a better place to live in than we can give our children." Scenes of children playing in dirty alleys follow, along with views of shacks, outhouses, women pumping water from outdoor pumps, and most striking of all, sickly looking, apathetic children staring vacantly at their unhealthy surroundings.

The movie shows a skyscraper, then water in the harbor. People eat standing up, on street corners; everyone is mechanized and appears to be in a hurry.

Finally the camera pans to the city of the future, with huge dams, airplanes, and trains—technology being harnessed for the good of individuals. The commentator states dramatically, "This new age builds a better kind of city, close to the soil once more. New cities take shape, green cities, they're built in the countryside, they're ringed with fields and trees and gardens. The new city is organized to make cooperation possible between machines and men and nature." A view of Greenbelt from above is on the screen as the commentator continues: "The sun and air and open green a part of the design, quiet neighborhoods are built into the pattern and built to stay there."

Views of Radburn and Greenhills are shown while the commentator begins to describe the new towns:

> This is no suburb where the lucky people play at living in the country. This kind of city spells cooperation. Wherever doing things together means cheapness or efficiency or better living. Each house is grouped with other houses close to the school, the public meeting hall, the movies and the market. Around these communities a belt of public land preserves their shape forever. The children need the earth for playing and growing. Bringing the city into the country. Bringing the parks and gardens back into the city. Never letting the cities grow too big to manage. Never pushing the meadows, fields, and woods too far away. This works as well for modern living as once it did in old New England towns.

By the end of this description the viewer is looking at scenes of Greenbelt, which continue until nearly the close of the film. Children ride bikes on a tree-lined sidewalk, dropping under the street on a pedestrian underpass, while tranquil music plays; the sky is blue, with fluffy white clouds. Other children play baseball, swim in the lake, play in a playground. A quick scene shows the interior of one of the houses with its sleek modern furniture and appliances.

The camera follows a paperboy on his paper route, past houses, the community center, the shopping center, playgrounds, while the commentator intones,

> The people who laid out this place didn't forget that what we need are air and sun for growing, whether it's for flowers or babies. . . . You can't tell where the playing ends and the work begins. We mix them here. We learn by living. Playgrounds, schools, libraries are meant for everyone. Not just for a few who get the breaks. . . . These new communities rest foursquare on men and women, their need for friendly faces, their need to have a public world that's just as human as their private homes.

The final scenes of the movie flash back to the slum housing and then forward to scenes of Greenbelt, while the commentator concludes: "You take your choice. Each one is real and each one is possible. Shall we sink deeper in old grooves, paying blight with human misery? Or have we vision, have we courage? Shall we build and rebuild our cities clean again, close to the earth, open to the sky? Can we afford a house, a neighborhood, a city, as good as this for everyone? . . . You take your choice. Each one is real, each one is possible. . . . We've got the skill. . . . We've built the city."[5]

Presented here is the story of Greenbelt, Mumford's ideal made real.

# Greenbelt, Maryland

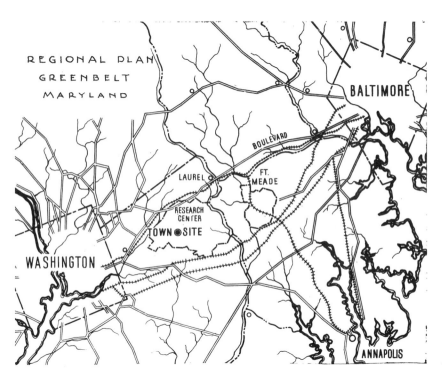

Greenbelt, Washington, D.C., and Baltimore. Farm Security Administration, Prints and Photographs Division, Library of Congress

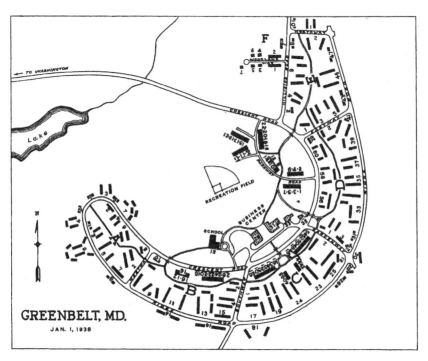

Greenbelt, Maryland, January 1, 1938 (showing superblocks
A–E, interior walkways, pedestrian underpasses, shopping
center, and school). Artist's rendering by J. Norvell, courtesy
Tugwell Room, Greenbelt Public Library

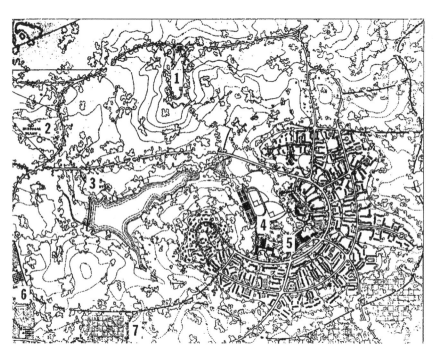

Hale Walker's plan for Greenbelt, 1936 (1. water tower,
2. disposal plant and incinerator, 3. picnic center and lake,
4. community center, 5. store group, 6. rural homesteads,
7. allotment gardens). *Architectural Record*, September
1936, p. 219

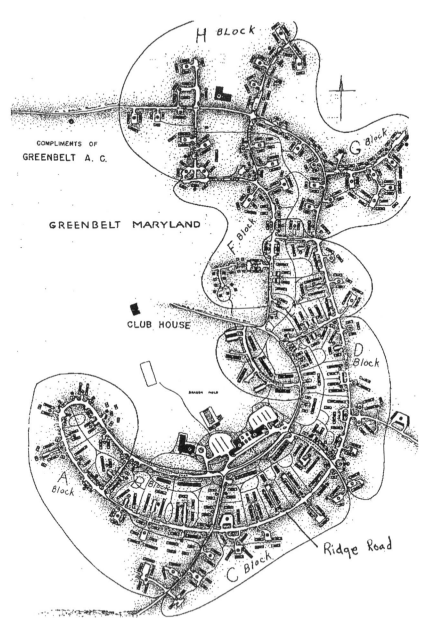

Greenbelt after addition of defense homes, drawn by L Grenahan, September 1944 (original superblocks were A–E; row houses were added outside of Ridge Road and blocks F, G, and H, north of Northway). Courtesy *Greenbelt News Review*

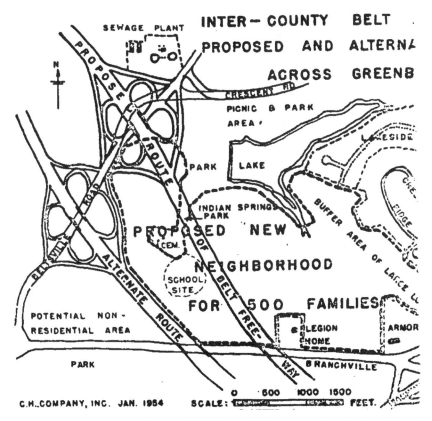

Proposed Washington Beltway route through Greenbelt and
the alternate route that was actually selected, January 1954
(proposed route was rejected because it went through a
cemetery). *Greenbelt Cooperator,* January 21, 1954

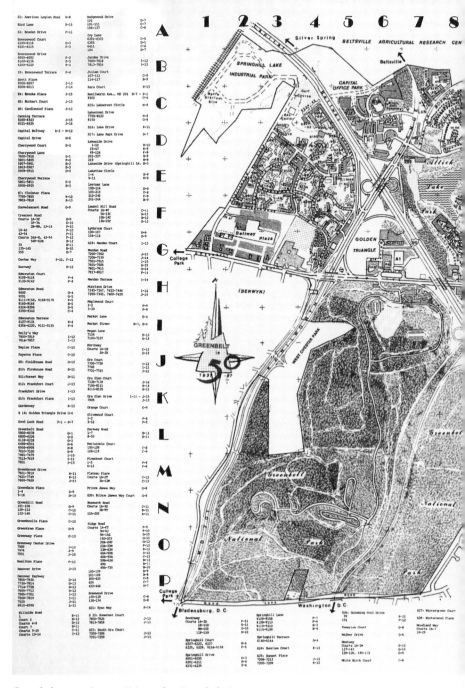

Greenbelt, 1987. © 1987 Citizens for Greenbelt, Inc.
Drawn by Map Committee, Cliff Brown, chairman

# I

# Greenbelt's Founding

# 1

# Building a Planned Community

THE TOWN OF GREENBELT displays its ideals on the façade of the building that serves as its community center and school. The preamble to the United States Constitution, in the form of an art deco frieze by the sculptor Lenore Thomas, illustrates "We the people," "in order to form a more perfect union," "establish justice," "ensure domestic tranquility," "provide for the common defense," and "promote the general welfare." These goals, boldly proclaimed by the Founding Fathers, also set forth the hopes of those planning a new community, who dared to dream of a better life for all residents. Among those daring to dream of a better life was Rexford G. Tugwell, a member of Franklin Roosevelt's Brain trust and the major proponent of New Deal programs to build planned communities.

Greenbelt in fact owes its existence to Rexford Tugwell, who left his position as professor of economics at Columbia University, where he had been especially interested in agricultural problems, to help Roosevelt devise a plan for economic recovery during his 1932 presidential campaign. Immediately after Roosevelt's inauguration, the president appointed Tugwell assistant secretary of agriculture. Tugwell assisted in formulating the National Industrial Recovery Act and supported the Department of Agriculture's attempt to strengthen food and drug law enforcement (which never received a vote in Congress after strong lobbying by food and drug companies). He advocated control over industrial profits and the use of social planning to bring about an economic recovery.[1]

Tugwell persuaded Roosevelt to create the Resettlement Administration in the spring of 1935, using money set aside for unemployment.[2] The Resettlement Administration consisted of existing programs transferred from other federal agencies and consolidated into a single organization. The Subsistence Homesteads Division of the Department of Interior, the Rural Rehabilitation Division and the Land Utilization Division, both from the Federal Emergency Relief Administration, and the Land Policy Section of the Agricultural Adjustment Administration formed the core of the Resettlement Administration. The greenbelt town program, administratively under the Suburban Division, formed the only new program of the Resettlement Administration.[3]

The Resettlement Administration, while receiving much media attention and criticism, was only one among many of Roosevelt's New Deal programs. The Emergency Relief Appropriation Act of 1935, which provided the $126,500,000 funding for Resettlement Administration programs, including $31 million for the greenbelt town program, totaled $4,889 million, most of which went to agencies such as the Works Progress Administration and the Public Works Administration. The Resettlement Administration received no further funding, although in 1938, the government allotted a further $3,750 million to relief agencies. The greenbelt town program thus received only a small portion of the federal funds expended during the New Deal.[4]

Tugwell saw three purposes for the Resettlement Administration, the first being the provision of financial aid for marginal farms through small loans, with the addition of technical and education aid programs as needed. The second purpose was to move rural families from unproductive land to more fertile areas, as part of a land reform program. The third purpose was to build suburban towns, providing housing for poor urban dwellers outside city centers, where cheap land made the project economically feasible.[5] In addition the greenbelt town program would provide jobs for workers during the construction phase. Tugwell hoped to build several dozen green towns, but political and financial realities forced the number to three: Greenbelt, outside of Washington, D.C.; Greenhills, Ohio, outside of Cincinnati; and Greendale, Wisconsin, on the outskirts of Milwaukee.

Tugwell described the origin of the program in his diary notes of March 3, 1935: "FDR let me off city housing, though he laughed at me for not wanting to do it. I talked to him about satellite cities as an alternative and interested

Transient workers clearing land, Greenbelt, November 1935.
Photo by Carl M. Mydans, Farm Security Administration,
Prints and Photographs Division, Library of Congress

him greatly. My idea is to go just outside centers of population, pick up cheap land, build a whole community and entice people into it. Then go back into cities and tear down slums and make parks of them. I could do this with good heart and he now wants me to."[6]

As a planned community, Greenbelt had several components: a physical design based on Clarence Stein's plan for Radburn, New Jersey; the sociologist Clarence Perry's neighborhood unit; and a social plan of economic and social cooperation. Clarence Stein, in *Toward New Towns for America,* provided a succinct analysis of the key design elements, his own Radburn plan, and the neighborhood unit—and how they were utilized in the Greenbelt project. Stein felt that elements of his Radburn design realized their fullest potential in Greenbelt, including blocks with central greens, separation of automobile and pedestrian traffic, cul-de-sacs, pedestrian underpasses, and homes facing their garden side, with their backs facing the street.[7] As the illustrations demonstrate, the use of Stein's ideas helped give Greenbelt its unique look.

The concept of the neighborhood unit figured prominently in Greenbelt's physical plan. The sociologist Clarence Perry, who worked with Clarence Stein and others during the 1920s in the loosely knit Regional Planning Association of America (RPAA), stressed that his ideas provided for the development of a sense of community. Perry first published his work as part of the Regional Plan of New York and Environs, sponsored by the Russell Sage Foundation. His neighborhood unit consisted of an elementary school and a community center in a building centrally located and within walking distance of an entire residential cluster. Other centrally located buildings might be churches, shops, a library. The neighborhood boundaries were major arterial streets, but within the neighborhood the roads carried local traffic only.[8]

Greenbelt's plan clearly followed Perry's ideas, as Stein commented:

> The three [greenbelt] towns are among the best applications of the principles laid down by Clarence Perry, which we would have carried out at Radburn had its growth not been stunted. Each Greenbelt Town in the beginning was, in effect, a single neighborhood. The focus of each is a planned neighborhood center consisting of school, community buildings, shopping center, government and management offices, and principal recreation activities. They are each built around such a planned center.[9]

The idea of combining physical and social aspects of community planning came from the RPAA. This group, formed in 1923, consisted of architects, planners, and social critics rebelling against current development. They sought regional cities, the preservation of small towns and villages, and the reconstruction and renewal of urban centers. The group served as an informal school for training and education in community planning. Led by Clarence Stein, members sought alternatives to what they perceived as shoddy, crowded housing. According to Roy Lubove, in his history of the RPAA, "The kernel of the RPAA's program was the cooperation of the social architect and planner in the design of large-scale group and community housing, financed in some measure by low-interest government loans, and directed toward the creation of the regional city."[10]

The RPAA reflected Stein's own political and social beliefs. Influenced by the Ethical Culture Society, he desired active reform to achieve a more equitable society.[11] Working to make his ideals a reality, Stein created Sunnyside, New York, built at Queens, Long Island, with the landscape architect Henry Wright. Alexander Bing, a real estate developer, assisted by organizing the limited-dividend City Housing Corporation in 1924 to finance land acquisi-

Row house construction, Greenbelt, March 1936. Photo by
Carl M. Mydans, Farm Security Administration, Prints and
Photographs Division, Library of Congress

tion and construction. At Sunnyside Stein and Wright first experimented
with the organization of blocks to create common green space at the center of
each block. Wright had previously designed a block plan, demonstrating that
housing density could be maintained even with land allotted for green space.

In 1928, the City Housing Corporation purchased two square miles in
Fairlawn, New Jersey, where Stein and Wright created an innovative plan to
solve the problem of "how to live with the auto." The formation of Radburn
included residential blocks in which homes faced a central green space, their
backs toward the streets; cul-de-sacs within the block, with all major roads on
the periphery; and a complete pedestrian walkway system inside the blocks,
with underpasses between blocks. The collapse of the U.S. economy in 1929
caused the bankruptcy of the City Housing Corporation before Radburn
could be completed.[12] However, the designs reappeared, to be used in the
greenbelt towns; Lewis Mumford commented, "Had it not been for the ideas
that the Regional Planning Association . . . had put into circulation during

President Franklin Roosevelt and Resettlement
Administration head Rexford Tugwell (*to the president's
right*), inspection tour during construction, Greenbelt,
November 1936. Photo by Arthur Rothstein, Farm Security
Administration, Prints and Photographs Division,
Library of Congress

the twenties, the Greenbelt Towns undertaken by the Resettlement Administration in 1935 would have been inconceivable."[13]

The layout of Greenbelt, in accordance with RPAA ideas combining the
physical and social aspects of planning, encouraged the formation of friendships. Children tended to play with others in their row of townhouses. Townhouses facing each other formed a court, whose residents frequently formed
social groups. Court residents typically had Halloween parties and picnics,
and they played softball games against residents of other courts. Friendships
formed among court families remained prominent in the memories of early
residents.[14] Courts thus became the smallest physical unit for organizing in
Greenbelt. For example, the Kindergarten Committee carried out a survey,
by courts, to determine if interest in a nursery school existed.[15]

The next largest unit by which residents organized themselves was the block, each consisting of fourteen to eighteen acres and containing approximately 120 homes facing in toward a wooded area. The town consisted of five blocks (A, B, C, D, and E), each formed of courts connected by walking paths and bounded by streets. Individual blocks sponsored consumer discussion groups in May 1938.[16] The Cooperative Organizing Committee, in its effort to enlist subscribers to the cooperatives, sponsored a series of block nights in April 1939, during which representatives of the cooperatives answered questions from residents.[17] Greenbelt organizations thus evolved ways of utilizing courts and blocks to break the population into small groups. The largest social unit was created by the school. Center School contained kindergarten through eighth grade, and since most families had at least one child in school it served as a unifying element.

While the design of Greenbelt reflected the influence of Stein, a planner, and Perry, a sociologist, the third aspect central to Greenbelt's plan, social and economic cooperation, received its impetus from Tugwell. Tugwell hoped the towns in the greenbelt program would embody his dream of places where people could live and work together cooperatively. The historian Paul Conkin, in *Tomorrow a New World,* says that Tugwell's "desire for a collectivized, co-operative society was almost a religion."[18] The timing proved auspicious for Tugwell, as cooperatives formed a large part of New Deal programs in agencies such as the Agricultural Adjustment Administration, the Farm Credit Administration, the Tennessee Valley Authority, and the Rural Electrification Administration. According to the historian Joseph Knapp, "It was a time when consumer cooperatives were buoyed up by enthusiasm and great hopes. Cooperation was touted as 'the Middle Way,' the alternative to Communism or Fascism."[19]

Tugwell's own interests showed clearly at a planning meeting he organized to formulate the Resettlement Administration, held at the mountain resort of Buck Hill Falls, Pennsylvania, from June 30 to July 3, 1935. Tugwell saw his work in very broad terms, as he explained in his letter of invitation to Eleanor Roosevelt: "I am eager that from the beginning this housing should be thought of not only in terms of its physical aspects, but as a contribution to a better way of living." Mrs. Roosevelt attended the conference, as did Bernard Baruch, John Dewey, prominent planners and architects such as Clarence Stein and Catherine Bauer, and many officials of the new Resettlement Administration. Items on the conference agenda included, first, "Social objectives of resettlement projects and methods of evaluating progress," second, "Selection of residents to go into the projects," third, "Responsibilities and duties of management," fourth, "Training for project managers," and fifth, "Statement of physical equipment and facilities required for the social program." Social objectives received primary attention, while facilities ranked last on the agenda, revealing their relative importance to Tugwell.[20]

Soon after the Buck Hill conference, the planning and design staffs began their work. On December 12, 1935, Tugwell described current progress to the president: "The most interesting thing is to watch the town and site planners work. The top salary we can pay an architect or planner is $5,600. In spite of that we have the best in the country and sometimes almost their whole staffs at work. . . . They work all hours, often all night, sometimes 36 hours at a stretch. But out of it there are gradually growing four complete communities of which I think you may be proud."[21] Tugwell hired John Lansill, a friend since college days at the University of Pennsylvania, to head the Suburban Resettlement Division, which was responsible for all three greenbelt towns. Frederick Bigger had the job of chief planner for all three towns; Hale Walker was town planner for Greenbelt; Wallace Richards was overall coordinator for Greenbelt; the principal architect for Greenbelt was R. J. Wadsworth.[22]

The "modern" buildings they produced in Greenbelt could best be described as plain and down-to-earth, as judged by both contemporary observers and historians. An advisor to the project, Henry Churchill, called the home exteriors "competent and undistinguished." The historian Joseph Arnold labeled them "functional" or "contemporary," while urban planner David Myhra categorized them as international style.[23]

Although the houses were plain, the planners achieved some variety through use of building materials. Some homes were constructed of cinderblock; others were frame and covered with either brick or mineral siding. Different floor plans provided additional variety: for the row homes alone there were seventy-one plans. Units came with one, two, or three bedrooms; some had combined kitchen and dining rooms, while others had two separate rooms. Most units were two stories, but sixteen were one story and twenty-two had three stories, with walkout basements.[24] The total housing in Greenbelt consisted of 885 units: 574 row houses, mainly two story; 306 units in four-story apartment buildings; and 5 prefabricated detached homes.[25] Hale Walker sited Greenbelt's housing in a crescent shape along a wooded ridge, aligned so the open end of the crescent faced prevailing summer breezes.

Ideas soon to become common in construction were first tried out at

Lenore Thomas and school–community center frieze
depicting the preamble to the U.S. Constitution.
Photo courtesy Greenbelt Homes, Inc.

Greenbelt, such as threaded copper fittings for water pipes, brass plumbing in the waste system, and decorative glass blocks, used in the school and the façades of the apartment buildings.[26] The Suburban Division aimed to produce structures with low maintenance and replacement costs, so high-quality items were used in the construction. This followed Clarence Stein's advice in a 1935 study done for the Resettlement Administration in which Stein assumed that the residents would be low-income families, so operating costs would need to be low.[27]

Home interiors were small but functional. Artists working under the direction of the Special Skills Division of the Resettlement Administration designed furnishings especially for Greenbelt's dwellings. Manufacturers selected by the Procurement Division built the furniture, which was sturdy, simple, and elegant.[28]

In each greenbelt town questionnaires that the Suburban Division sent to local area residents helped determine what community facilities would be included in the center. These families expressed a desire for a library, a swimming pool and other recreational facilities, a theater, and a community hall.[29] Planners created a small shopping center, which contained a central open space for informal gatherings as well as a movie theater. Behind this were the swimming pool and ball fields, next to it was the school, which served as the community center and also housed a library. The lake was also available for recreational purposes.

Landscaping proved an important aspect in a town with *green* in its name. Under the direction of Angus MacGregor, the clearing of land during the construction phase yielded up to 13,000 plants, shrubs, and small trees. These were transplanted to a nursery and later used to landscape the grounds. Workers planted many of FDR's favorite tree, the tulip poplar, and hedges were used instead of fences to divide yards.[30]

The building of Greenbelt took place in an isolated area of Prince George's County thirteen miles northeast of Washington, D.C. The land had already been optioned for federal government use by the Federal Emergency Relief Administration and the Subsistence Homesteads Division of the Interior Department. It was transferred to the Resettlement Administration along with these programs.[31] Since Tugwell wished to start the greenbelt town program quickly, he decided to utilize land already belonging to the federal government.

Tugwell moved too slowly for George E. Allen, head of the District of Columbia Transient Bureau, who precipitately announced that the Greenbelt project would be able to employ all the area's unemployed workers. Officials of the Resettlement Administration had planned to establish contact with county and state officials to gain their cooperation with Resettlement Admin-istration plans before construction began, but Allen's announcement super-seded theirs. In a meeting with Prince George's County commissioners, a U.S. senator from Maryland, Lansdale Sasscer, revealed local fears "that the town would in time become the harboring place for disreputable characters from all along the Atlantic seaboard."[32] After administration officials con-vinced local and state authorities that the workers would be closely super-vised and the future tenants of good character, they received reluctant ap-proval of the project.

Due to George Allen's eagerness to put his men to work, the construction of Greenbelt became a race between the architects and the construction

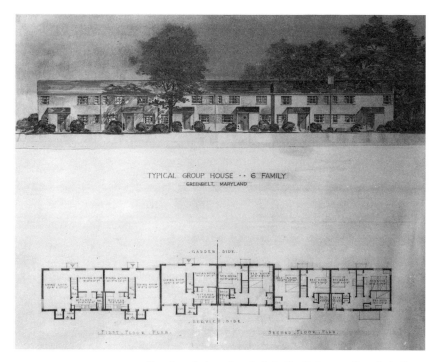

Sketch and plan of typical six-family house, Greenbelt. Farm
Security Administration, Prints and Photographs Division,
Library of Congress

engineers. As soon as the architects drew their plans they were blueprinted
and rushed to the construction site. Since the workmen arrived at Greenbelt
two weeks before even the street plans were available, they were given make-
work such as digging the lake with picks and shovels. Throughout construction workers used almost no labor-saving machinery, as a main point of the
project was to create jobs.[33]

Many nearby residents in Prince George's County remained unimpressed
by official sanction of these workers and did not approve of the project.
Washington-area newspapers editorialized against the construction, reflecting local sentiment and inflaming it further. The *Washington Times* declared
in an editorial entitled "Who Is Forgotten?": "Some of the folks living in the
area near the proposed town see the project as a nuisance and a dangerous
nuisance at that. The Prince Georgians, apparently, were not to be given any
chance to stop the town building before it got started."[34] The *Washington Post*
stated that destitute families from drought areas would be housed in Green-

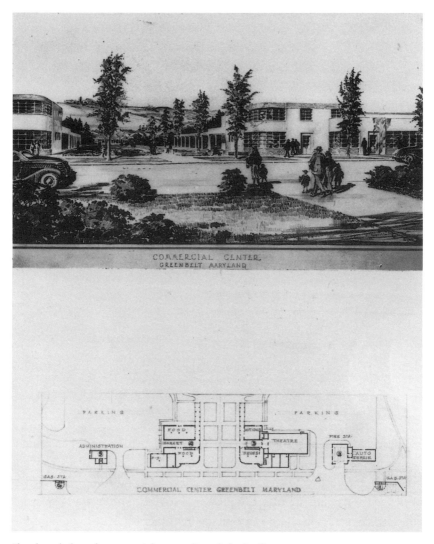

Sketch and plan of commercial center, Greenbelt, April 1936.
Farm Security Administration, Prints and Photographs
Division, Library of Congress

belt. Although it was not true, the statement exacerbated local feeling against construction. Prince George's County residents also resented the importation of transient labor from Washington when many of their own citizens were unemployed.[35] According to the director of the Prince George's County Historical Society, people resented the new town for two reasons. First, the inhabitants would be "foreigners," meaning anyone not from Prince George's

County. Second, Greenbelt residents would be provided with many services that others in the county lacked, such as paved roads, a swimming pool, recreation programs, kindergarten, and nursery school.[36]

Life in Prince George's County in 1937 provides a key to understanding local reaction. It was largely rural, and small towns such as Laurel and College Park existed only along Route 1. Very little growth had occurred from the late nineteenth century to the time Greenbelt burst onto the scene. In the 1930s Prince George's County had the lowest percentage of registered voters in Maryland, as residents largely left their civic affairs in the hands of Senator Sasscer and his political machine. Prince George's County residents clung to a rural, unchanging way of life, accepting the status quo in which most had no real opportunity for active participation in civic affairs. Greenbelt provided a great contrast, as a willingness to participate in civic activity was required for tenancy.[37] The obvious disapproval of their neighbors helped create in Greenbelt residents a siege mentality, which reinforced their physical isolation.

Throughout 1936, the first year of construction, it became apparent that the prominence of Rexford Tugwell had both positive and negative effects on the greenbelt town program. He knew many people of influence, and Roosevelt trusted him. He could present his ideas to Congress, state and local officials, and the public, using the authority of the Roosevelt administration to see that programs took shape as he wished. Thus he used his influence for the greenbelt town program, gained from Roosevelt's public backing, to see jobs created and low-cost housing built, all the while pushing his ideas of economic and social "cooperation." However, throughout 1935 and 1936 Tugwell began to turn into a political liability for the president. One problem stemmed from Tugwell's manner and the way he presented himself. His academic background and way of thinking presented difficulties in his relations with the public, as described by the historian Bernard Sternsher in his *Rexford Tugwell and the New Deal*: "There is no doubt that Tugwell's writings and speeches presented a knotty problem in communication. Although he was capable of plain talk, his academic writings were often sophisticated, given to sweeping concepts, and abstruse."[38]

Newspaper editors not well disposed toward New Deal programs caused another problem for Tugwell by misrepresenting the Resettlement Administration. A typical example occurred in November 1935 when the *New York Times* ran an article on its front page headed, "Tugwell Has Staff of 12,089 to Create 5,012 Relief Jobs." Tugwell, in a reply to the editor, stated that his

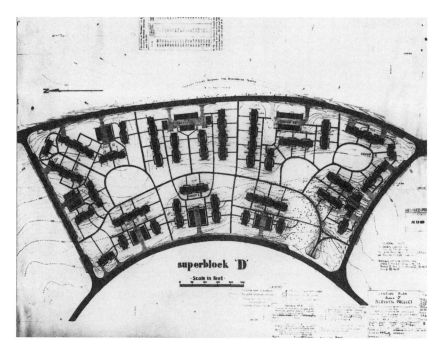

Plan of a Greenbelt superblock. Farm Security
Administration, Prints and Photographs Division,
Library of Congress

agency cared for 354,000 farm families, or about 1,500,000 persons. Articles
in the *Washington Post* and the *Baltimore Sun* accused Tugwell of overspend-
ing, objected to the $5,500,000 allowed for the construction of Greenbelt, and
ignored the employment aspect of the program. Tugwell replied that construction costs would indeed be high, as one purpose of the project was the
provision of jobs for 2,000 workers on relief. All work, even digging the lake,
was done by hand, to allow for the maximum possible number of laborers—
which, of course, increased the cost. Even though Tugwell responded to such
articles as best he could, the negative publicity continued to mount.[39]

Commentary on Resettlement Administration programs reveals the diversity of national sentiment toward the greenbelt town program and, indeed,
toward the New Deal itself. Greenbelt, more than any other Resettlement
Administration program, received intense scrutiny throughout the period of
its construction, partly due to its proximity to the national press corps. Detractors nicknamed Greenbelt "Tugwell's Folly" and began belittling the town
before it left the drawing board.[40] Liberals in favor of the administration

of President Roosevelt and his New Deal programs approved of the greenbelt towns. Conservatives, including residents of Prince George's County
and anyone against New Deal policies, detested the greenbelt towns. All
three Washington papers—the *Post*, the *Evening Star*, and the *Times-Herald*—
took a conservative stance, publishing frequent editorials against Roosevelt,
against the New Deal in general, and against Greenbelt in particular. Eugene
Meyer, the governor of the Federal Reserve Board under Herbert Hoover,
subsequently purchased the *Washington Post* and editorialized against "Tugwell's Folly" at every opportunity. Because administration officials and congressmen read these newspapers daily, the papers had great influence.

In the opposing camp, the liberal publication the *New Republic* heralded
the greenbelt towns "as object lessons in wise community planning." In a 1936
*New Republic* article, planner Henry Churchill stated,

Aerial view of Greenbelt, November 1937. Farm Security
Administration, Prints and Photographs Division,
Library of Congress

These towns represent a great advance over anything hitherto done. They will be four towns where decent housing and the amenities of life are permanently provided for, where amenities for people, rather than profits, were a primary consideration. That has never happened before. . . . This country is large enough and wealthy enough to afford such an experiment. The prevailing philosophy of self-liquidation, of constipated conservatism, must not be allowed to interfere with what is, by any philosophy, next to the T.V.A. the most significant of the New Deal's attempts to be a New Deal.[41]

Labor unions gave the greenbelt town program a hearty endorsement. The *Trades Unionist,* the official organ of the Central Labor Union of Washington, D.C., carried a resolution of support in its June 6, 1936, issue. Publications that editorialized in favor of New Deal programs—such as the *Literary Digest, Life,* and the *New York Times Magazine*—emphasized the positive aspects of Greenbelt.[42]

Some coverage was factual and neutral in tone, such as *Business Week's* series of articles describing the progress of the cooperative stores. The *National Municipal Review* published a detailed examination of Greenbelt by Cedric Larson, covering the history, plan, utilities, charter, and city management. The *Christian Science Monitor* described Greenbelt in a brisk, matter-of-fact manner. The *Washington Star* published a series of five articles that conveyed both the pluses and the minuses of the new community. Professional journals such as the *Architectural Record* carried detailed descriptions of the greenbelt towns that emphasized design rather than ideology.[43] These publications demonstrated that it was possible to discuss the greenbelt town program from a nonpolitical stance. However, the majority of coverage criticized New Deal programs in general and Tugwell and the greenbelt program in particular, often expressing outrage at the "communism" or "socialism" of the towns. Those in private enterprise who felt threatened by the greenbelt town construction often led these attacks.

Some of Tugwell's difficulties stemmed from his willingness to deflect attacks from President Roosevelt. He had been serving in this role when he took charge of the Resettlement Administration in 1935. Thus "this choice of the administration's principal whipping boy guaranteed hostility toward the RA's program on the part of conservative Congressmen and a large sector of the press."[44] Tugwell's enthusiasm for cooperation along with his visibility made him an obvious target for those who disliked the more socialistic aspects of New Deal programs. He attempted to fight back through the press, writing articles for sympathetic publications such as the *New Republic* and

Gable-roofed houses and connecting paths, Greenbelt,
May–June 1942. Photo by Marjory Collins, U.S. Office of
War Information, Prints and Photographs Division,
Library of Congress

*Work,* published by the District of Columbia Works Progress Administration. He defended himself in speeches, trying to explain his programs.[45] Finally, in a vain attempt to halt criticism of Resettlement Administration programs, Tugwell resigned from his position on December 31, 1936.

Press attacks continued, however, until the future of the greenbelt program was in jeopardy. Constant harping by congressional critics responding to those in private enterprise who perceived a threat from greenbelt towns stimulated Roosevelt to dismantle the Resettlement Administration, transferring its remaining programs, including the greenbelt town program, to the Farm Security Administration (FSA) on January 1, 1937. The Construction Division of the FSA took over responsibility for the completion of Greenbelt, under the direction of army engineers.

Wallace Richards, the coordinator in charge of Greenbelt for the Resettlement Administration, retained an active interest in the town even though he had been removed from his job with the end of the Resettlement Administra-

Brick-veneer row house, Greenbelt, September 1936. Photo
by Drier, Farm Security Administration, Prints and
Photographs Division, Library of Congress

tion. He was dismayed at the way army engineers planned to finish the town.
Having met President and Mrs. Roosevelt when he conducted them through
town on their tours of inspection, and perceiving their interest—particularly
hers—he wrote to Mrs. Roosevelt to express his deep concern:

> Engineers, and in particular army engineers, too often are aware of only the imme-
> diate job to be done, and are apt to feel no responsibility for the future social
> significance of their work. Last minute economies made by them nearly always are
> achieved in two ways,—at the expense of educational and recreational facilities.
> Thus, drastic curtailments at Greenbelt will mean minimum landscaping, inade-
> quate exterior painting, no playground equipment, no allotment gardens, no town
> forest, no regional park and playground, and a lake which would not be developed
> for recreation,—in other words, last minute economy would affect those very
> things which we had counted on to hold the townsfolk at home, make them happy
> and community conscious, and prevent them from spending a disproportionate
> amount of their income in seeking relaxation elsewhere.[46]

Richards realized that the heart of the town—what would make it a special
place—would likely be destroyed before it came into being. He added, "If I

were not utterly convinced of the necessity for the full completion of Greenbelt, I would not think of begging you, as I am, to assume an even greater interest in the town." Mrs. Roosevelt scribbled this note on the letter: "FDR: This seems to me most important. There is a new experiment being involved which curtailment may seriously hurt. Won't you look into it? E.R."

On November 6 Mrs. Roosevelt received this memo from her husband: "I suggest that you and Will Alexander go out to Greenbelt and see it some morning or afternoon after you get back. F.D.R."[47] On December 9 Mrs. Roosevelt and Will Alexander, now head of the greenbelt project, journeyed to the town.[48] Her visit received favorable coverage by the media; she herself gave it publicity by describing it in her syndicated column, "My Day." Two days later other distinguished visitors toured Greenbelt: Representative Henry Ellenbogen and Henrietta Klotz (secretary to Henry Morganthau Jr., who was secretary of the treasury); they were accompanied by Reginald Wadsworth, principal architect, and W. F. Baxter, press relations chief of the FSA.[49] Perhaps because of this demonstration of interest by persons in positions of power and influence, Congress appropriated the necessary funds, and the town reached completion as originally planned.

Throughout the period of construction, Greenbelt's notoriety, achieved by massive press coverage, created great interest in the project on the part of Washington area residents. The *Washington Daily News* reported on March 10, 1937, that "more than 350,000 persons have visited Greenbelt, Maryland, since officials began keeping count." In June 1937 the city manager, Roy Braden, reported that he had received thousands of inquiries regarding rental without ever sending any literature announcing availability. In the greater Washington area, from which all applicants would be drawn, word-of-mouth advertising sufficed to attract record numbers of prospective tenants.

On September 2, 1937, the Department of Agriculture issued a lengthy press release, "Farm Security Administration Announces Greenbelt Rentals," which explained who was financially eligible and how to apply. The FSA organized carefully to handle the expected deluge, preparing maps showing the exact location of Greenbelt and, within it, the Tenant Selection Office. These were distributed throughout federal offices in Washington. FSA officials hired a staff of thirty to interview and select residents from the more than 5,700 families who applied to live in the 885 units available.[50]

The FSA used several criteria to select the 885 families. Income limits eliminated many prospective residents. The Resettlement Administration originally stated that Greenbelt residents would be "low-income" workers but later altered the norm to "modest income." The income range allowable for a family of four was between $1,100 and $2,000 and for a family of five between $1,200 and $2,100. Families with incomes $50 above or below these limits were accepted if they met the other guidelines. The small size of the units dictated that only families with six members or fewer could be accepted. Heads of families had to be at least twenty-one years of age; families with young children received preferential consideration.

The administration attempted to select people who would represent a cross section of the metropolitan Washington, D.C., region. They developed a quota system for place of residence, with 80 percent from the District of Columbia and 10 percent each from Virginia and Maryland, reflecting the regional population distribution. Quotas for place of work limited federal government workers to 50 percent, District of Columbia government employees to 5 percent, and nongovernment workers to 45 percent. Individual departments and agencies of the federal government even had quotas, to make sure any one employer, such as the State Department, the Department of Agriculture, or the Commerce Department, was not overrepresented.

To ensure against religious discrimination in selection, the administration developed quotas based on a religious census taken in 1926 showing the Washington area to be 59 percent Protestant, 34 percent Roman Catholic, and 7 percent Jewish. However, the first residents were overwhelmingly Protestant, with fewer than 34 percent of them Catholics.

Only whites lived in Greenbelt, which has occasioned some criticism. (Several black families did live on farms on federal land surrounding the town but played no part in town life.) However, in 1937 America a racially integrated community would have been not only an anomaly but totally unacceptable to the citizens of Prince George's County. Greenbelt's original plan called for a separate area, the Rossville Rural Development, to be built for Negro families. It would have included almost one-third of the Greenbelt tract. Officials quietly dropped the plan and later explained that Negroes had their own low-cost housing project in northeast Washington, Langston Terrace, built by the Public Works Administration. The greenbelt town program certainly did nothing to ameliorate racism in housing or to help blacks achieve better housing.[51]

In addition to the above restrictions, other, more amorphous qualities had to be met to become a resident of Greenbelt. The original selection

Gable-roofed houses and their garages, Greenbelt,
May–June 1942. Photo by Marjorie Collins, U.S. Office of
War Information, Prints and Photographs Division,
Library of Congress

plan stated: "The family should desire to meet objectives of the community, namely to raise their standard of living by taking advantage of the improved living conditions offered; as well as to participate in a cooperative-minded community for the mutual advantage of the group both from the economic and social standpoints."[52] To make decisions on these factors, a staff investigator conducted a home interview, determining if a genuine need existed for adequate housing and if the family seemed economically stable. FSA staff obtained credit data on prospective residents as well as information from employers and landlords.

The staff also assessed the willingness of the family to "cooperate and participate in activities of the town." The second page of the questionnaire used by the interviewers concerned "Community Factors." Under the heading "Understanding of Project," the interviewer could mark a scale ranging from "Little Understanding of Project and No Evidence of Ability to Adjust" to "Good Understanding of Project, Ability to Adjust, Definite Interest in

School–community center, Greenbelt, September 1937.
Photo by Arthur Rothstein, Farm Security Administration,
Prints and Photographs Division, Library of Congress

Participation." The interviewer also marked down the extent of the family members' participation in organizations in the past and their "hobbies and talents." Under the heading "Desire for Project," the scale ran from "Indifferent or One Parent Does not Wish to Move to the Project," to "All Members of Family Eager to Go to the Project." The last item, "Family Integration," provided the interviewer with the choices of "Questionable Family Life and Social Attitudes," "Evidence of Lack of Family Solidarity or Evidence of Social Problems in Family," and "Well Integrated Family Group, Normal, Home Loving, Self-Respecting." Since so many families applied, presumably only the ones marked at the highest end of the scale on community factors would have been selected.[53] Who was actually responsible for selecting the tenant families, and the process used, are not known.

In 1937 FSA officials selected families that fit all the above guidelines to become the original tenants. However, after a reduction in FSA staff in 1938, tenant selection had to be accomplished in a less time-consuming manner, eliminating the home interview. The criteria remained size of income, size of family, credit data, age, and income stability.

Given the selection process, it would seem likely that Greenbelt residents would be a fairly homogeneous group. However, this proved to be untrue, as a division into conservative and liberal camps existed from the beginning and pervaded town life. The liberal majority was composed of families with young children, with parents in their late twenties; the conservative minority was older—children in their teens and fathers often veterans of World War I. The liberals formed cooperatives and civic associations, while the conservatives threw their energy into the development of the largest American Legion chapter on the East Coast. The conservatives, as well as the liberals, must have convinced FSA interviewers of their intent to abide by the cooperative ideology. Once in Greenbelt, they had the freedom to pursue cooperation in the manner they desired.[54]

The actual construction of Greenbelt began on October 12, 1935. The first tenants moved in on September 30, 1937, when the first housing blocks reached completion. Families occupied each block of units as soon as the workers left, the last being completed in September 1938.[55] By the end of November 1937 about 150 families had moved in; by the next April, 483; and by June, 610. Thus the town filled up gradually over a period of a year.

As the number of people in town grew, the need for some form of administration became obvious. FSA officials took the steps necessary to put it in place. First, they wrote a town charter, which they took to the state assembly for approval. On June 1, 1937, the Maryland General Assembly obligingly passed a law just for Greenbelt authorizing the first council-manager form of government in the state. The plan called for a nonpartisan five-person city council to be elected by residents for two-year terms. Council members, in turn, would select a mayor from among themselves.[56]

The FSA hired Roy Braden as the first city manager. As administrator, he determined the rules to maintain the safety and appearance of the town, and

A court of cinder block houses, Greenbelt, July 1937. Photo by Arthur Rothstein, Farm Security Administration, Prints and Photographs Division, Library of Congress

his first requirements were typical of those anywhere: residents had to keep their homes and yards orderly, not make loud noises, and park in their proper parking spaces. In an effort to promote cooperation he tried to govern by consensus. He asked residents what rules they felt would be necessary and used their responses to formulate guidelines. He also kept in mind Greenbelt's role as a federal government showcase and the large number of visitors who still turned up, especially on weekends. This led to a somewhat controversial rule limiting when laundry could be hung to dry, especially on Sunday.[57] FSA officials met Greenbelt's safety needs by hiring a director of public safety, who then created police and fire departments. By 1938 Greenbelt had a police force of three men as well as a functioning volunteer fire department.[58]

In spite of their overall approval, some early tenants had criticisms of their new planned environment. Most of them strongly disliked the isolation of the community, which caused a number of problems. Residents declared both the shopping facilities and transportation to Washington inadequate. Women criticized the design of the units, especially the small kitchens and the difficulty of doing laundry.

Other elements of the design received almost universal approval, especially the system of street underpasses and paths, which greatly improved safety for children. Original Greenbelters felt most strongly about aspects of the physical plan that affected their relations with others. They praised the extensive recreation program, with swimming pool, ball fields, equipment, and a staff ready to organize and assist.

The neighborhood unit plan, with community amenities focused in the heart of town, created a center that became an important aspect of Greenbelters' lives. By "the Center" residents meant the elementary school, which also functioned as a community center, and the adjacent area, which housed shops as well as an open space with benches. Descriptions varied, but all

FACING PAGE:
*Top*: Living room of row house, Greenbelt, December 1936 (furniture designed for Greenbelt housing by artists in Special Skills Division of the Resettlement Administration). Photo by Arthur Rothstein, Farm Security Administration, Prints and Photographs Division, Library of Congress
*Bottom*: Bedroom of row house, Greenbelt, November 1936 (furniture designed for Greenbelt housing by artists in Special Skills Division of the Resettlement Administration). Photo by Arthur Rothstein, Farm Security Administration, Prints and Photographs Division, Library of Congress

agreed on the importance of the Center: "The auditorium at the Center School was a truly multipurpose room. The high school basketball games were played there, our high school proms and dances were held there, graduations, concerts, town fairs, high school plays, political meetings—even the Community Church met there every Sunday for many years. It was truly a community center." Another resident remembered: "It was a very, very active town. You'd go to the Center and there were always people there talking, discussing the affairs of the city. . . . You could just feel the vitality."[59] July Fourth, Labor Day, and other holiday celebrations took place at the Center.

In spite of the abrupt end of the Resettlement Administration, the innovative physical plan of Greenbelt became a reality in the form originally envisioned through the special efforts of Wallace Richards, Eleanor Roosevelt, and Will Alexander. The design's courts and blocks encouraged the formation of social groups. With housing units clustered in a semicircle around the Center, all residents could easily walk to the school, the community center, and the shopping area. Thus as hoped the physical design contributed to the ability of residents to carry out their goal of cooperation by providing ways for residents to easily come together. The many ways they did so are discussed in chapter 2.

FACING PAGE:
*Top*: Kitchen in one-room apartment, Greenbelt, February 1938. Photo by Russell Lee, Farm Security Administration, Prints and Photographs Division, Library of Congress
*Bottom*: Greenbelt resident reading *Greenbelt Cooperator*, May–June 1942. Photo by Marjory Collins, U.S. Office of War Information, Prints and Photographs Division, Library of Congress

# 2

# Creating a
# Cooperative Community

IN NOVEMBER 1937, with enough homes and families in place, the city manager, Roy Braden, undertook the economic and social cooperative aspects of Greenbelt. He explained the approach his staff would use: "We realize that we have a job of education on our hands. We aren't going to force anything on the families. We will help to develop and further group living as contrasted with an individual philosophy."[1] This job of education must have been aided by the fact that the families had been selected in light of their stated willingness to cooperate. A number of the pioneer residents thus felt great personal responsibility for making the cooperative experiment of Greenbelt a success.[2]

There is no doubt that the townspeople evinced a willingness to organize. During the first year they formed thirty-five organizations. Some appealed to specific age groups, genders, interests, and political orientation. Other organizations were designed to bring diverse people together, such as the Citizens Association, the Junior Citizens Association, and the Civic Forum. Groups appealing to specific interests were the Journalistic Club, American Legion Post 136, Mother's Club, Boy Scouts, Parent-Teacher's Association, Bridge Club, Greenbelt Players, Camera Club, Girl Scouts, Men's Athletic Association, Brownies, Garden Club, Singles Club, and Community Band.[3] Residents kept themselves busy, as illustrated by "The Calendar of Events" published weekly in the *Cooperator*, which was begun in November 1937 to help keep residents abreast of local activity. The calendar typically listed more than forty meetings or events scheduled each week.

The goal of cooperation was only partially fulfilled by the formation of social organizations, because economic cooperation remained an integral part of the Greenbelt plan. The founding of economic cooperatives got off to an early start when, in August 1937, the Consumer Distribution Corporation (CDC, founded by Boston department store merchant Edward Filene to establish cooperative enterprise) leased the business center of Greenbelt from the government. The CDC, financially underwritten by Filene, provided not only funds for—but also advice on—the formation of business cooperatives. An early resident, attracted to the first organizational meeting by a handbill left on his doorstep, provided an account for the newspaper on Greenbelt's introduction to the economic aspects of cooperation, which, as he described it, had a social component as well. Fifty attended the meeting: "I did not know a person present but I listened to Messrs. Herbert E. Evans and Flint Garrison, members of the CDC Board of Directors. They explained that it was their desire to surrender ownership of Greenbelt business enterprises to us. When I realized they really meant it, I began to take notice of those about me and to hope I could know them; I hoped that we might be friends and work toward the realization of this dream."[4]

Interested Greenbelters formed the Cooperative Organizing Committee to decide what cooperatives they should create. The first year they developed a credit union, gas station, supermarket, drugstore, barbershop, and theater. In November 1938 the Cooperative Organizing Committee began to sell shares in the co-op businesses to town residents by going door to door throughout the community. Greenbelt Consumer Services, Incorporated, became a legal entity and went into business by January 1940, when shareholders elected the first board of directors. Early residents agreed that those most active in the formation of co-ops were the few who moved to Greenbelt specifically for its planned co-ops as well as liberals who agreed with the co-op nonprofit philosophy.[5]

The cooperative idea extended beyond formation of businesses to the provision of other community services needed in the isolation of rural Prince George's County. Greenbelt had no resident physicians and Prince George's County had no hospitals, so to meet medical needs, residents formed the

*Mother and Child* statue, shopping center square, Greenbelt,
May 1942. Photo by Marjory Collins, U.S. Office of War
Information, Prints and Photographs Division,
Library of Congress

Greenbelt Health Association in January 1938 along the lines of the Group
Health Plan that had begun in Washington, D.C., the previous year. The
Health Association solicited members throughout the community, who paid
a monthly fee for care. The association doctor could also treat nonmembers,
including non-Greenbelt residents, for a fee. The Health Center opened on
April 1, 1938, and by May 1, 1939, a small hospital began operation in the only
available space, several converted row houses; it had beds for six adults and
six children.[6] In a similar manner, interested parents formed a kindergarten,
as the state of Maryland did not provide education before first grade. Other

residents worked informally with town officials to set up a public library and bus transportation between Greenbelt and Berwyn, where a streetcar line ran to downtown Washington.[7]

One cooperative, called the Journalistic Club, formed to meet the urgent need for communication within the community. Volume 1, number 1, of the weekly paper of the club, christened the *Greenbelt Cooperator,* appeared on November 24, 1937. The Journalistic Club planned to publish six issues—which were typewritten and mimeographed—and then evaluate the situation. In the first issue they enunciated its policies:

1. To serve as a nonprofit enterprise.
2. To remain nonpartisan in politics.
3. To remain neutral in religious matters.
4. To print news accurately and regularly.
5. To make its pages an open forum for civic affairs.
6. To develop a staff of volunteer writers.
7. To create a good neighbor spirit, promote friendship, advance the common good, and develop a Greenbelt philosophy of life.

Greenbelt lake, September 1938. Photo by Marion Post Wolcott, Farm Security Administration, Prints and Photographs Division, Library of Congress

The paper published notices of upcoming meetings of the town's many committees and clubs and reported on their activities. After the initial six-week period, the Journalistic Club decided to create a more formal organization, becoming the Greenbelt Cooperative Publishing Association. The association began accepting advertisements for the *Cooperator* and charged readers 5 cents an issue. Through the months of 1938 new sections of the paper appeared: sports; reports on the many activities of the Recreation Association; an expanded woman's page called "Mrs. Greenbelt"; reports on cooperatives in other towns; "Greenbelt Junior," which was by and about children; and "Neighborhood News." In March 1938 the paper became a producer cooperative, with 5 percent of the profit divided among those who created it. In September the first printed paper appeared, along with the first photographs. The search for advertising succeeded, as virtually every co-op business in town placed advertisements in the paper. As a result, the staff was able to provide the paper free to every resident. It was delivered by the Boy Scouts. Copies were sent to President and Mrs. Roosevelt as well as to others involved in Greenbelt's formation, such as Rexford Tugwell.[8]

After one year of publication, the newspaper staff took stock of their situation and evaluated their progress. In the first anniversary issue, November 3, 1938, an editorial again listed the paper's objectives and judged whether they had been met. The editorial concluded that indeed "these policies have been pursued consistently by each successive staff of the *Cooperator*." The paper clearly served its purpose as a neutral vehicle of communication for the town. The staff took especially seriously policy 7, "To create a good neighbor spirit, promote friendship, advance the common good, and develop a Greenbelt philosophy of life."

Toward this end, in March 1938 the paper sponsored a contest for a Greenbelt flag and seal. The contest attracted entries from sixty-five residents. Mary Clare Bonham, a high school junior, won the flag contest for her design of a green pine tree on a white field bordered by two broad green stripes. Mrs. Robert Templeman created the seal: a map of Greenbelt with Greenbelt Lake included.[9] To remind readers of Greenbelt's beginning, the newspaper staff put out a special Charter Day edition on June 1, 1938, marking the official beginning of Greenbelt as a legal entity in the state of Maryland one year before. To emphasize the beliefs they wished to promote, the staff filled small leftover spaces in columns with blurbs such as "Be loyal to your community" and "Patronize the Greenbelt stores."

Greenbelt swimming pool, June 1939. Photo by Marion Post
Wolcott, Farm Security Administration, Prints and
Photographs Division, Library of Congress

In January 1940 the paper's staff began a continuing series titled "Our Town," giving Greenbelt's history, emphasizing its beginning as a cooperative planned community, and touching on everything from the Revolutionary War patriot buried in a local plot to the location and operation of the disposal plant. Thus even when the town was only two years and three months old, its unique beginning as a cooperative planned community served as a focal point for all that followed.

The success of the "Greenbelt philosophy of life" depended on citizen involvement, so the newspaper devoted much column space to encouraging activism by residents. The following June 22, 1938, editorial is worth quoting in its entirety because it served as a model, its call for communal participation repeated endlessly by the paper's staff:

> The Citizens Association has been reporting out of its committees splendid programs of community activities. It is evident that the Association, along with the Administration, is providing us with distinguished leadership.
>
> Now it is up to all of us to determine by our participation in these programs the measure of satisfaction they will give us.
>
> It is now, more than ever, up to us to say whether Greenbelt is to be a community of social significance or merely a group of cheap houses.

Co-op gas station, Greenbelt, September 1938. Photo by
Marion Post Wolcott, Farm Security Administration, Prints
and Photographs Division, Library of Congress

An editorial on November 10, 1938, called "Opportunity and Responsibility," offered similar commentary: "The people of Greenbelt are the pos-
sessors of opportunities enjoyed by few in history. To them has been en-
trusted the task of molding the policies of a new town. Here, as always,
Opportunity grasps the hand of his twin brother, Responsibility." The editorial went on to urge citizens to attend the meeting of the Cooperative
Organizing Committee the next Wednesday to decide if the town would take
over the management of the cooperatives. The exhortation proved effective,
as residents turned out to create Greenbelt Consumer Services.

On May 18, 1939 the staff published an editorial entitled "Greenbelt Citizens Must Make Their Fair." After explaining the idea of the town fair, the
editorial states: "What is required then is response, a world of response, from
every group, club, block, and row of houses. Citizens of Greenbelt: What do
you want in your fair?" On some occasions the newspaper staff did not have
to prod the citizenry. Individuals wrote letters to the editor such as this one in
the January 4, 1940, issue: "Do you remember when you moved to Greenbelt
and for the first time caught the true spirit of our community? There was the
hope of a better life, of better surroundings, of adequate shelter, and of

opportunity to be of service to your fellow men." The writer urged his readers to remember and benefit from the "cooperative philosophy."

FSA officials hoped to use a cooperative philosophy to meet the religious needs of the residents. In response to suggestions by a group of interested residents and FSA officials, the Office of the Chief Engineer of the FSA drafted plans, dated April 1, 1940, labeled "Proposed Religious Group, Greenbelt, Maryland." The plans show a first floor with two wings that house Protestant and Catholic chapels, with smaller rooms labeled Mormon chapel and Hebrew synagogue. The chapels and the synagogue were two stories high. Adjacent to the Protestant chapel were a nursery, a community room, and two Sunday school rooms. The second floor contained more Sunday school rooms. The basement plan shows a kitchen, a fellowship hall with a stage, a Hebrew educational room below the synagogue, and Catholic education and committee rooms below the Catholic chapel. The building plans clearly expressed the hope of people in Greenbelt to meet the religious needs of the community while, at the same time, stressing the ability of various faiths to cooperate by being housed together. The plans did not reach fru-

Lunch counter in co-op drugstore, Greenbelt. Photo courtesy Greenbelt Museum

ition, however, as the beginning of the war stopped all non-defense-related construction.[10] When building resumed after the war, individual denominations erected their own churches, to the disappointment of those holding most strongly to the cooperative ideal.

Throughout the prewar years religious groups struggled to meet the needs of the faithful. With help from the Council of Churches in the national capital area, interdenominational church services began in Center School on November 4, 1937, leading to the formation of Greenbelt Community Church, a Protestant group. Roman Catholics could attend services at Holy Redeemer in Berwyn but soon began making plans for their own church, meeting in the movie theater in the meantime.[11] In January 1939 the sixty-two Jewish families in town were invited to the first organizational meeting of the Greenbelt Hebrew Congregation. By November 1939 Protestants, Catholics, Jews, and Mormons were holding regular services in the town.[12]

The only major societal function not left to residents to organize cooperatively was the school, which was operated by the federal government. FSA officials hired the staff for Center School, which housed all grades the first year, and decreed that it would be a "model of progressive education." In practice, this meant that the children participated in everything, even in the planning of the curriculum. The FSA brought in a professor from Columbia University to help the teachers learn the tenets of progressive education and how to implement them.[13] One of the first students clearly remembered "learning by doing." This included such activities as dancing to music of Shostakovich to "express yourself," building nature trails, visiting the co-op stores, and running a post office. Students created the first co-op in town, the Gum Drop Co-op. Special teachers of art and music provided instruction daily. There were no books and no homework. In the small classes each child received much individual attention.[14] The use of the progressive educational model could have contributed to Greenbelt's image as a "radical" town.

After the first year Center School included just first through eighth grades, due to the completion of Greenbelt High School, which opened in September 1938. Built by the federal government and run by Prince George's County, the new high school included many students who had previously attended Hyattsville High School and who came from the surrounding towns of Branchville, Beltsville, Berwyn, and College Park. Greenbelt High School maintained a traditional curriculum, which created some difficulties in adjustment for students coming from Center School. Some who went on to Greenbelt High

Co-op drugstore, Greenbelt. Photo courtesy
Greenbelt Museum

felt they had a considerable amount of catching up to do on factual material. However, they realized they had learned quite well how to analyze data and complete their own projects. Teachers at the high school felt that students from outside of Greenbelt resented the Greenbelt children because of all the special facilities and activities they had available to them.[15]

The children of Greenbelt, and adults too, could indeed, throughout the prewar years, choose from many town activities. In addition to clubs, school activities, and religious events, groups such as the Citizens Association actively planned events for the entire community. For the Fourth of July celebration in 1938 the association sponsored a parade, concession stands, contests, races, a community sing, and fireworks. An estimated 5,000 people participated, although the population of Greenbelt at the time was only 2,540. A May Day celebration in front of the school attracted a crowd; and Labor Day festivities that year included a parade, games, contests, and a dance. On December 22, 1938, the Citizens Association sponsored a Christmas party for more than 1,000 children. The group received much assistance from Roy Braden and his staff in assembling these massive community affairs.[16]

The creation of such community events formed part of a conscious tactic by the city staff to create a sense of community. Doctoral candidate William

Co-op grocery store, Greenbelt, February 1938. Photo by
Russell Lee, Farm Security Administration, Prints and
Photographs Division, Library of Congress

Form interviewed town officials in 1940 and described how they worked with
residents to create a sense of community:

> Efforts were made from the first year to create precedents that might become "tradi-
> tions." A Citizens Association was sponsored to promote widespread interest in the
> community. Whenever the organization flagged, the Administration made efforts
> to keep it alive or to reorganize it. A town flag was designed and displayed on proper
> occasions. On holidays the "town common" was highly decorated with the national,
> state and town flags, as well as with laurel, or other appropriate paraphernalia. The
> Administration actively sought to introduce organizations which were not but
> "should be" represented in the town; for example, the Boy Scouts. Financial support
> was given to "projects" which might become community traditions. The most
> conspicuous example of this was the Town Fair, [for] which the council was urged to
> appropriate rather large sums of money. The labor of town workers was used to
> build booths and prepare exhibits. Such a project when repeated two or three years
> became a "tradition" or a "custom." The Administration helped initiate programs
> for Memorial Day, Flag Day, Armistice Day, and other holidays. Although these were
> sponsored by private organizations, such expenses as printing the programs were
> assumed by the town. Everyone was urged to attend the convocations and many
> pains were taken to omit no participant's name from the printed program.[17]

Town administrators clearly stressed both social organizations and the formation of traditions as ways to create a sense of community. They regarded a sense of community necessary to the success of the cooperative part of their enterprise.

Another important aspect of life in a cooperative community, emphasized by both town officials and the newspaper, involved participation in the local, nonpartisan, political process. The first city election in November 1937 stimulated great interest, with eighteen people running for five council positions. Since the council experienced difficulty in selecting a mayor from among themselves, they decided that in future elections whoever got the most votes for council would automatically become mayor. In the 1937 election almost every eligible voter in town actually voted. In the November 1938 election, as the newness of participation began to wear off, 43 percent of the eligible voters cast their ballots, causing the *Cooperator* to decry the lack of participation.[18]

The city council developed the city budget and decided the taxation rate. Residents provided much input on these questions, forcing the council to accommodate their views. Form commented that "various groups in Greenbelt, like the American Legion and the Athletic Club, have been able to manipulate the council for their own purposes in the guise of 'community welfare.'"[19] And even though town politics remained officially nonpartisan, Form found that 58 percent of residents identified themselves as Democrats, leftists, or liberals; 13 percent as conservatives or Republicans; and 34 percent as independent. These groups formed shifting allegiances, frequently disagreeing about the running of community organizations, with the majority prevailing after a sometimes acrimonious exchange of views.[20]

The Hatch Act ruling by the Civil Service Commission in August 1939, decreeing that federal employees could not participate in local politics, greatly affected Greenbelt as well as other towns in the Washington, D.C., area. Activists held a protest meeting in Greenbelt on August 29, circulating petitions to be sent to the attorney general. In October 1940 W. C. Hull, executive assistant at the U.S. Civil Service Commission, ruled that federal workers could participate in nonpartisan city politics.[21] This ruling made local government possible for Greenbelt, since the percentage of residents connected to the federal government increased greatly during the war years, and it would have been difficult to form a city government without their participation.

During its first several years Greenbelt attracted as much attention as it

Co-op nursery school, Greenbelt, May 1942. Photo by
Marjory Collins, U.S. Office of War Information, Prints and
Photographs Division, Library of Congress

had during construction. Visitors included Sir Raymond Unwin, the "father
of British city planning," a large group from the International Institute of
Architects who were meeting in Washington, and representatives from the
Chilean government and the Pan American Union who were studying housing. Numerous sightseers, ordinary citizens, continued to visit on the weekends.[22] Many of those who had been involved in the creation of the town
maintained an interest in it. For example, J. S. Lansill, director of the Suburban Division, wrote to the *Cooperator* (of March 23, 1938): "To read about
your various activities, your live interest in your schools, in your cooperative
stores and in other civic matters, to observe your neighborliness and your

pride in the health, recreational and safety features of Greenbelt, gives all of us, who were responsible for the planning, immeasurable satisfaction."

But although most visitors admired Greenbelt, the city's relationship with the press in its first several years continued to be adversarial, with notable exceptions. In February 1938 the *Christian Science Monitor* published a detailed description of life in Greenbelt that pointed out both advantages and disadvantages. The *Washington Post* ran a lengthy evaluation of the town at the end of one year entitled "Greenbelt, One Year Old, Pays Social Dividends as Test Tube City." An adjacent article was headlined: "All Visitors Remark Utopian Happiness." Two other Washington papers, the *Star* and the *Times-Herald,* continued attacks on the town, often by ridiculing it. A typical example occurred in March 1938. Roy Braden sent a five-page notice to all town residents explaining regulations for town life. He had waited five months after families began moving in to see what rules might be necessary and what the people themselves would want. The rules seem reasonable and fair, but the headline in the *Washington Star* cried out "Greenbelt Bans Display of Wash After 4 p.m. and on Sundays," with a smaller headline underneath proclaiming "Private Mutterings of 'Regimentation' Are Heard as New Rules Go into Effect—Tricycles Regulated."[23]

Although every resident certainly did not agree with every regulation of town life, many resented the way they were pictured in such articles—as automatons with every aspect of their lives programmed. As one resident pointed out in an editorial in the May 18, 1938, *Cooperator*: "It seems useless to point out to Washington newspapers that the regulations in question, as well as others mentioned in the editorial, were the result of requests made by many of our citizens . . . [and] that most of the regulations, and other regulations, are to be found in Washington apartments and in communities established by private enterprise." Another resident wrote in the December 15 issue: "If Washington newspapers would endeavor to see the significance of this town, and not use it as a political football, they would find that they have been scorning a development which holds more for the future of democracy than all the impassioned utterances that will ever decorate their papers. Democracy will grow not from your saying things, but from people doing things."

Greenbelt citizens, in their attempts to speak for themselves, resembled David in his battle with Goliath because Washington and Baltimore newspapers, like others of the period, had become big businesses, after a period of consolidation of smaller papers beginning in the late nineteenth century.

Center School, Greenbelt, May 1942. Photo by Marjorie
Collins, U.S. Office of War Information, Prints and
Photographs Division, Library of Congress

Publishers represented the interests of large corporations, the National Association of Manufacturers, bankers, and investors. They strongly disliked the
New Deal, based as it was on the premise that business had failed the country
and only extreme measures could return the nation to prosperity. Roosevelt
felt that 85 percent of publishers opposed the New Deal; he often could
charm reporters but failed to persuade newspaper owners of the virtues of his
programs. The economist Rexford Tugwell, with his admitted desire for
control over business in order to use comprehensive long-range planning,
stood no chance against a press defending business as usual.[24]

The location of Greenbelt so close to Washington made it an ideal target for columnists representing their editors' conservative views. Greenbelt residents had no voice speaking on their behalf and no way to fight back other than to send letters to the editor and to fume in disgust in the pages of the *Cooperator*. Greenbelters' defense of their town perhaps helped change the stance of one publisher, Eugene Meyer, owner of the *Washington Post*. After vilifying Greenbelt since 1935, Meyer dramatically changed the editorial position of his paper at the time of Greenbelt's first anniversary celebration. The *Washington Star* and the *Washington Times-Herald*, on the other hand, continued to criticize Greenbelt.

The extensive press coverage may well have reinforced the opinion of Greenbelt residents that they were unique, a part of a historic experiment in cooperative living. From the beginning, articles in the *Cooperator* focused on the special nature of Greenbelt and emphasized each individual's role in its success. Perhaps this feeling of taking part in history contributed to the residents' not only saving documents of the early years but also recording their impressions of life in their new town, both of which are fortunately available to us.[25]

When residents described Greenbelt's most important aspect, they focused on the social climate, frequently described as a "small-town atmosphere." A typical comment described it this way: "It was like an old-fashioned small New England town where everyone took part and everyone felt a sense of responsibility." Some focused on economic equality: "There was no hierarchy here, no rich people looking down on the peasants. We were all equal." Several individuals remembered favorably the prevalent religious tolerance, which also helped create a spirit of unity.[26]

Greenbelt residents of the early years frequently called themselves pioneers. The first settlers lived in an isolated area, surrounded by fairly hostile natives, so the comparison seemed apt. A typical resident remembered: "It was really like a frontier community where you didn't bring in your ancestry and your D.A.R. connections. You contributed and you were taken for what you were, at face value." Another said, "When we were threatened from the outside, we would gather together, circle the wagons, and defend ourselves."[27]

The first tenants commonly described their social interrelations in the town with the statement, "We were like one big family." Individuals using this phrase sometimes meant all of original Greenbelt, sometimes just the families in a particular row or court. As one woman who grew up in Greenbelt explained: "We had such a feeling of security, always. We knew wherever we

Center School high school graduating class, Greenbelt, 1938
(this was the school's only graduating class; after 1938 Center
School housed only grades 1 through 8). Photo courtesy
Greenbelt Museum

wandered, playing, somebody's mother would be looking out for us." Another
representative comment: "We cared for, and shared, each other's families."[28]

Greenbelt residents and former residents utilized comparisons to a pioneer town or a family to explain what made Greenbelt unique: the strong
sense of community. This sense of communal identity was the single most
important characteristic of Greenbelt, as the following quotations attest.

> There was a feeling of sharing, of togetherness in Greenbelt that I'll never forget.
>
> Growing up in a small town where everybody knew you and cared, gave me a
> sense of belonging, and a true understanding of friendship.
>
> When I first came here, I was given a basket of food from the co-op, that really
> impressed me. I was given the local newspaper, information about all the civic
> associations, the schools, churches, telephone numbers. I was made to feel part of
> the family. . . . I think I was strongly attached to Greenbelt because of the closeness,
> the family atmosphere that we have here. People care.
>
> You could meet 100 Greenbelters anywhere, and 99 of them would say they
> never had it so good as they did in Greenbelt.[29]

This sense of community, in turn, created a sense of responsibility. As one early resident explained: "Somehow the sense that the community is everyone's responsibility permeated Greenbelt Elementary. No one littered our community any more than we would our own homes."[30] Many residents attended meetings to discuss community issues, sparking lively debate, as one who had often been in the center of the action described: "We would be going at it, hot and heavy about something. When it came time to vote, if the vote was close, we would line up along opposite walls so we could be counted. If we were on opposing sides from someone on one issue, it didn't matter. We knew we might be fighting together on the next one."[31] While residents frequently disagreed on exactly how they should run their town, almost all felt that their actions mattered, that they had a significant responsibility to fulfill in creating their cooperative community.

The labor of residents working together to create their community affected each individually. A portion of a letter by a town resident and co-op activist, Walter R. Volckhausen, published in the *Washington Evening Star* on December 13, 1938, provides an excellent summary of the impact of Greenbelt on people's lives: "We in Greenbelt have learned that, though as individuals we are feeble, as a group we have power. We have learned the significance and potentiality of united social action—and what greater lesson must our people learn if our democracy is to survive?"

Thus the planners of Greenbelt, who hoped to create a town in which cooperation and a sense of community existed, achieved their goal. The creation of community involved a number of factors, some intended by the planners and some not. Factors that were part of the initial plan were the physical design, especially the neighborhood unit, the careful selection of initial residents, thus providing many more activists than usual in a community, and the initial leadership provided by government and co-op businesspeople. Individuals working together in Greenbelt organizations, whether business, religious, or social, reinforced cooperation and helped create a sense of community, as did the creation of city traditions.

Other factors were unplanned. To keep costs low, the town had to be built on cheap land found only on the outskirts of cities, which caused the residents, in their isolation, to band together to provide necessary services and meet their social needs. Feelings of homogeneity, particularly as to income, helped them to see themselves as members of a group. Being attacked by the outside world, whether by newspaper editors or unfriendly Prince George's County residents, united them in defense of their community.

The goals of economic and social cooperation had thus been conveyed from Rexford Tugwell to those directly in charge of creating Greenbelt, such as Wallace Richards and his associates; then to those who took over the program in the Farm Security Administration, Will Alexander and his staff; and then to Roy Braden and other city officials to whom the task fell of making the ideas real. Last, the tenants themselves, working diligently, created their own model town in the wilderness, pioneers participating in the New Deal. By the end of 1940 Greenbelt residents felt quite pleased with themselves and their community. They had formed a viable town that met their needs and created a sense of accomplishment—as well as the hope that as their many organizations became well established, they would not have to work quite so diligently. They could reflect on their achievements only a short while, however, before the disruptions caused by war began to affect Greenbelt directly.

# II

# Change and Continuity

# 3

# The War Years

CHANGES CAUSED BY WORLD WAR II challenged Greenbelt residents and their newly developed way of life. By 1941 families in Greenbelt had created organizations to suit their needs, ranging from business co-ops to athletic groups, social and hobby clubs to religious organizations. Isolated in Prince George's County, thrown together constantly in their community work as well as informally in daily life, Greenbelt residents formed close friendships, forging an identity as Greenbelters, as they began to call themselves.

At the same time the federal government decided to build housing for defense workers in Greenbelt. One thousand units quickly appeared adjacent to the existing 885 units. Adjustment to the doubling in size of the housing area—and of the original tenants to the invasion of newcomers and of the defense workers and their families to life in Greenbelt—required change. Could a lifestyle of cooperation survive an influx of temporary newcomers larger in size than the number of original residents?

Looking at the makeup of Greenbelt's population before the wartime changes is helpful in understanding what followed. William Form's 1940–42 study of Greenbelt residents reveals the results of the Farm Security Administration (FSA) selection process.[1] Greenbelt consisted largely of young people (with a median age of 24.7 years). The proportion of children under 5 years of age was thus higher than that in the Washington, D.C., area, while older married couples with adolescent children were not as well represented as in the general population. Ninety-seven percent of Greenbelt residents were born in the area, compared to 66 percent of District of Columbia residents. Sixty-four percent of Greenbelt residents had completed high school and had some college education, while only 41 percent of District residents had a high school diploma. The figure for District of Columbia whites with a high school education was 49 percent. Thus Greenbelt residents had a considerably higher education level than the area average.

In 1940 Greenbelt reflected fairly closely the figures aimed at by the quota system of the initial selection process. Greenbelt was a predominately Protestant community; 25 percent of residents were Catholic, and 7 percent were Jewish. Less than 5 percent of workers were self-employed, and fewer than one-quarter did manual labor. Clerical workers composed 56 percent of the workforce, most of them federal government employees; 14 percent of workers were categorized as professional or semiprofessional. Greenbelt could thus be summarized as a community of white, Protestant, low-level government employees.

Data collected by Form for the years 1939–42 show that "the advance of Greenbelters up the income ladder for the three-year period was quite rapid." The increase in government employees in wartime Washington gave these individuals a boost up the career ladder, as did their relatively high educational level. Greenbelt wage earners during this time "easily surpassed the middle income range in the country."[2] Seventy percent of Greenbelt's white-collar workers intended to get more education, mainly to aid in career advancement. Almost 60 percent of Greenbelt fathers intended to give their children a college education. In 1943 and 1944, 45 percent of Greenbelt high school graduates intended to go to college. (In actuality, a number ended up in the armed services.) At this time, only 7 percent of Maryland high school graduates attended college. Thus a picture emerges of a community of families with many shared characteristics. They valued education and appeared upwardly mobile, advancing in income due to the rapid wartime expansion of the federal government.[3]

These residents received a flood of newcomers to Greenbelt as part of war preparation by Congress, which passed the Lanham Act in February 1941, authorizing construction of homes for defense workers. Greenbelt's 1,000-unit project was among the largest of the forty-three housing projects across

Defense worker housing under construction, Greenbelt,
November 1941. Photo by John Vachon, Farm Security
Administration, Prints and Photographs Division,
Library of Congress

the country. The town's selection as one of the first sites arose due to the existence of vacant government land. Of the 3,411 acres purchased by the federal government to construct Greenbelt, only 120 acres was used for the existing 885 dwellings. Utilities were already in place to serve additional homes, since the original plan called for gradual expansion to 3,000 units. Thus using Greenbelt as a site saved time and money for the federal government. The completed units came under the authority of the Federal Public Housing Authority (FPHA), which was created in 1942 as a war housing authority as part of the consolidation of federal housing agencies ordered by President Roosevelt. The FPHA assumed control of all programs previously under the aegis of the FSA, including original Greenbelt.[4]

The building of the new homes in the north end, around their own elementary school (not completed until after the war) created another neighborhood unit according to Clarence Perry's definition. It had been hoped that his plan, which had been utilized in the original design, would lend itself to the creation of a cohesive unit within its boundaries. In this case, the physical design contributed to the division of Greenbelt into two groups centered on the two elementary schools. Although the town was physically divided into two distinct parts, the effort of fighting the war united the town.

Construction of the new homes by the Federal Works Agency (FWA) began immediately after the Lanham Act passed; they were built in the same styles and patterns as the original dwellings, thus blending into the community. However, construction went along at a hectic pace and with materials significantly inferior to those in the original homes. The frame construction buildings had asbestos shingle siding and "rubberoid" roofing. Unlike the first section, the houses lacked garages, the courts had no inner walkways, and no pedestrian underpasses channeled children safely under Crescent Road. The area received only minimal landscaping, giving it a barren look.[5]

The FWA eliminated as unnecessary the finishing touches that Wallace Richards and Will Alexander successfully fought to retain in the original homes, thus deviating from the physical and cultural meaning of Greenbelt, where the pedestrian underpasses and inner walkways contributed crucially to the social goal of cooperation. Underpasses and walkways provided easy access for residents to each other's homes and to the Center, encouraging both formal meetings at the Center and informal gatherings in the courts. Furthermore, the FWA proceeded very slowly in the laying of sod, and most yards in the new section were seas of mud.

The FWA deviated not only from the design but also from the original FSA guidelines for choosing tenants, who no longer had to go through the rigorous selection process endured by the first residents of Greenbelt. The FSA had already removed the quotas for religion, place of residence, and place of employment but maintained the income, family size, credit, and age criteria. In contrast, the FWA criteria for occupation of a wartime housing unit was simply employment by the war or navy department and an income below $2,600 a year.[6]

Families began moving in immediately upon completion of the first homes in December 1941, rapidly expanding the town's population from 3,200 to 8,000 and creating the largest community in Prince George's County. (See table A.1 for Greenbelt's population, 1940–90.) The war required an adjustment not only to a rapid size increase but also to continuous population turnover. According to the Family Selection Office, from September 1941 to September 1942, a 33 percent turnover in the population of original Greenbelt had occurred. Such change had previously been about 20 percent a year.

The numerous population shifts caused turmoil in community groups, as many of those heading organizations—who had resided in Greenbelt from the beginning—left town. Between January 1940 and September 1941 four men served as mayor, and councilmen had to be continually replaced when they were transferred elsewhere by the government. The city manager Roy Braden, who provided some stability throughout this period, also left in October 1943; he was replaced by James T. Gobbel. Other community institutions affected were the *Cooperator,* which had nine editors between May 1940 and July 1945, and Greenbelt Consumer Services, which constantly had to replace people on its board of directors.[7]

Even while coping with flux in community groups, Greenbelters attempted to use their organizations to welcome the newcomers. In May 1942 Greenbelt Consumer Services held a get acquainted party for newcomers. The Citizens Association had a dance the following June for the same purpose. In August the Citizens Association formed a system of block captains to aid new residents in becoming accustomed to town life. Town traditions such as special events on holidays continued but in a low-key manner.[8]

The newspaper played a key role in welcoming newcomers, but an editorial that proposed holding meetings for new residents met with no response. After the newspaper repeated the suggestion, the Citizens Association executive committee approved the idea and met with leaders of other groups in town to formulate plans for meetings to welcome new residents. In March 1942 a weekly "Hi, Neighbor" column made its appearance, in part to welcome new tenants. In April another new column, called "I Want to Know . . ." answered questions sent in by readers regarding town life and services. Throughout 1942 and 1943 articles on the history of community organizations such as the *Cooperator,* Greenbelt Consumer Services, and the Health Association were published for the benefit of new residents. In September 1941 a feature called "One Year Ago" appeared, and on November 20, 1942, a series called "Five Years Ago" began, to "give new residents insight and perspective on some of the town's present problems." It was also meant to provide "older residents pleasant memories of the days when everything was new and exciting."

As part of its continuing effort to integrate the newcomers, a December 26, 1941, editorial headlined "Welcome Neighbor" began, "You are the first of a thousand families who are moving in next door or across the street, and we want to get acquainted as friends and neighbors. We want you to like

Newly completed defense housing, Greenbelt, December
1941. Photo by Arthur Rothstein, Farm Security
Administration, Prints and Photographs Division,
Library of Congress

what is good in us and help us correct our faults, which are numerous and
apparent." The article then listed the fears of the community:

> We confess that you frighten us just a little, because you—a thousand families of
> you—can completely upset the direction of our community efforts if you wish. You
> can outvote us, turn the town away from cooperative enterprise, smash the civic
> organizations we have labored so hard to form, undermine our community pride,
> and indeed turn Greenbelt topsy-turvy if that is your will, for we shall soon be only
> a minority in the town we so jealously regarded as our own little province.

The editorial counseled newcomers and old-timers on desirable behavior:
"Here we both have a problem that will require our best efforts. We must
dismiss from our minds any shadow of word or action which might imply
that we regarded you as outsiders or interlopers. You will need to keep in
mind the greatness of your own power and be sure that it is used to the
common advantage of all of us."

The newspaper pointed out progress made toward the formation of a
unified community. On January 8, 1943, an editorial proclaimed, "The shib-
boleth of a divided town was discarded Monday night at the Citizens Association meeting when residents of the older FSA homes and the newer FWA
defense homes joined in voting action on community problems plaguing all
of us." Addressing the same issue on February 19, an editorial commented:
"There is some feeling that the people in the new defense area do not belong

to the community in the sense that residents of 'old' Greenbelt do. This is unfortunate, and does not represent the true situation."

Although the newspaper spoke in this vein, it continued to refer to "Old Greenbelt" and "defense homes." A July 10 article about a war casualty stated that "Captain Yeatts was very well known among the residents of Old Greenbelt." On August 20 an editorial commented: "The rather high feeling which developed between old and new Greenbelt when the newer section was first opened up is on a rapid decline, thank goodness. There are a few who still cling to a feeling of superiority, but they are rapidly coming around to a more sensible point of view." The editorial admitted that "perhaps the *Cooperator* has not done its part in helping to alleviate such feelings. In the future the paper will not make any reference to the different sections of Greenbelt as old or new or defense and non-defense. From this point on any home in Greenbelt is in Greenbelt and not in any particular part of town unless it is designated as north or south or center." An editorial in August 1945 noted that the new head of the Citizens Association was a "North-Ender," a term that still effectively demarcated the town. Evidently the newspaper staff felt it was not a pejorative label, as was "defense homes."

While Greenbelters tried to adjust to the changes, hoping to make the newcomers part of their community, they attempted to maintain the co-ops and other organizations that gave their community its identity, a goal especially difficult to achieve during wartime. An onerous period followed for the many co-op businesses, which had to deal with more than double the previous number of customers while at the same time losing many experienced board members and store staff. During the war the co-op businesses in Greenbelt's central shopping area consisted of a food store, a variety store, a theater, a valet shop, a drugstore, barber and beauty shops, a gas station, and a tobacco shop. Sales in 1941 totaled almost $450,000; in 1942 they reached $690,00; in 1943, $1 million.[9]

Co-op management made efforts to respond to the needs of residents, especially the new ones. Tenants in the defense homes protested that they had to walk too far to purchase groceries and other necessary items. Greenbelt Consumer Services did not have enough capital to erect another building but responded quickly when shoppers suggested adapting residences in the north end for a small store. This store began operation in January 1943 in four row houses.[10] Interest in the co-op remained high: ten candidates ran for four places on the board of directors in August 1944, and in December of that year more than 325 of the approximately 1,500 members came to the annual meeting. By the following January, the co-op reported 1,609 share-owning members, of a possible 1,885. The co-op board actively and successfully recruited members among the new residents, using weekly articles in the *Cooperator* to discuss the origin and function of the co-op, followed by exhortations to join.

The co-op idea continued to be used in areas other than businesses. A co-op nursery school began operation in April 1941. A child care center, which provided care from 7 A.M. to 6 P.M., opened in September 1943 in the first floor of several housing units. It was organized with funds provided by the Lanham Act. In August 1944 parents initiated a program for before- and after-school care. In December 1944 a gathering place for teenagers was provided by the city in the basement of the firehouse–police station. Teenagers cleaned and painted; boys in shop classes built benches and tables, while girls sewed slipcovers for the donated furniture. Called the Drop Inn, the facility was operated by its teenage members.[11] In October 1944, the Housewives Club was formed to "aid new mothers who are unable to obtain adequate care for home and child, the mothers who have no place to leave children while on shopping trips to the District, and housewives who find shopping difficult because of illness or other demands."[12]

Another new cooperative came into existence in September 1941 when the Journalistic Club reorganized itself into the Greenbelt Cooperative Publishing Association, Incorporated. The newspaper maintained a positive financial situation throughout the war due to the large number of advertisements placed by Greenbelt Consumer Services—as well as the occasional federal government pitch to join the navy. Although on a fairly firm financial footing, the newspaper remained critically short of help throughout the war, and new volunteers needed to be recruited continually to replace those leaving town. In January 1941 an article asking for volunteers appeared, apparently provoking little response, for the next week the staff offered a six-week course in journalism, free, to anyone willing to work on the paper during that period. This ploy proved successful, at least for a time.

On March 26, 1942, an editorial again pleaded: "Today we find ourselves at the end of our rope and must again appeal to you for aid. We need you, and we do mean you, if this paper is to continue." The editorial encouraged newcomers, in particular, to come forward: "We would like to see some of

View from shopping center roof, Greenbelt, May 1942.
Photo by Marjory Collins, U.S. Office of War Information,
Prints and Photographs Division, Library of Congress

our new residents on the paper, so don't be bashful. We all live here and we can all help to make this a better place to live if we cooperate in our efforts. There is no money in these jobs and we can assure you there is very little thanks or glory; just the satisfaction of knowing that you are working on a free press dedicated to the advancement of Greenbelt and the protection of the consumer." In spite of the high turnover in Greenbelt's population the *Cooperator* succeeded in producing weekly issues throughout the war years, continuing in its mission to serve the town by highlighting the past, serving current needs, and maintaining Greenbelt traditions. The April 3 issue sadly noted the disappearance of local groups: "Many organizations have passed from the local scene. The Greenbelt Players, the Better Buyers, the Hospital Auxiliary, Men's Glee Club, Hobby Club, and a dozen others quietly folded up and disappeared." Adjusting to meet the current interests of residents, the paper began the column "Greenbelters in Uniform" in June 1943.

Newspaper editorials broached the topic of the town fair persistently throughout the war. The second town fair had been held on Labor Day weekend in 1940, with Will Alexander, the former head of the FSA, and Hale Walker, the original site planner, both in attendance. The fair went on as usual in 1941, but as the *Cooperator* noted in July 1942, no plans were as yet under way, although organizing usually began in May. An editorial in August lamented the fact that the fair would not be held due to inaction by the city council. In April 1943 the *Cooperator* again brought up the idea of a town fair and asked, "Anybody interested?" The mayor announced at the council meeting the following week that the fair would not be held because people could not devote time to it during the war. On August 25, 1944, an editorial appeared lamenting the fate of the fair. It ended: "We are serving notice now that early next spring the *Cooperator* will start plugging for a Town Fair in 1945, to revive the best tradition of Greenbelt before it fades too far into dimming memory." This statement hints at what no one said explicitly, that the town fair was the tradition most enjoyed by residents and the one they most wished to see continue, being closely linked to the identity of the town.

In 1945 the newspaper's campaign began early, with an editorial in January proclaiming "On with the Fair." The city council responded by setting things in motion at a meeting in February. On February 9 the *Cooperator's* front-page headline observed: "Town Fair Movement Gets up Steam as Leaders Climb on Bandwagon." The August 24 edition proclaimed the result of many months' effort in a banner headline, "Come to the Fair!!" The fair had been successfully held once again.

The newspaper continued to cajole and bully townspeople into becoming "active citizens." Sometimes its voice became strident. On April 10, 1942, in an unusual front-page editorial (editorials were usually on the second of four pages) a list of things people should be doing and were not was followed by this question: "Just who in the Hell do you people think you are to live here off the fat of the land without so much as raising a finger to help with the vital civic functions so necessary in Greenbelt?"

The unique beginning of Greenbelt continued to be invoked not only by the newspaper but other town groups as well. A city directory published by the American Legion in 1943 provided a "brief outline of the construction and purpose of the town of Greenbelt." When Franklin Roosevelt died, town administrators conducted a memorial service at Center School, where the "auditorium was full to overflowing." James Gobbel "recalled the time when the President personally released the first batch of fish to stock the man-made Greenbelt lake in 1938, and the several times that he visited Greenbelt homes."[13] The *Cooperator* commented: "Our town is a living memorial to the spirit and the hopes of the great leader whose departure we mourn. It is our task to keep it a worthy memorial."

Greenbelters remained interested in improving their town, using direct action to obtain their goals. Transportation to the District of Columbia continued to be difficult, prompting a hundred people at a Citizens Association meeting in May 1942 to protest the Capital Transit Company's effort to reduce shuttle bus service. As a result of the protest, Capital Transit maintained service at current levels.[14] In June 1945 the tenants of one of the apartment houses wrote to officials of the FPHA complaining of lack of heat and hot water, an effort that produced action.[15] Interest in local government in Greenbelt remained high in the period preceding the war but declined somewhat during the war itself. In September 1941 fourteen candidates ran for five places on the city council; in September 1943 only six candidates vied for office. Interest revived quickly, however, as the war wound down. In 1944, 78 percent of Greenbelters eligible to vote went to the polls in the presidential election. As would be expected, voters in Greenbelt chose Roosevelt over Dewey, 244 to 161.[16]

During the war years, Greenbelt residents threw themselves with gusto into the "war effort." As early as December 1939 members of Greenbelt co-op stores voted not to allow German or Japanese goods to be sold in the stores. In May 1940 the *Cooperator* printed the opinions of twenty community leaders regarding the war. Most felt strongly that the United States should arm the Allies but not become more directly involved. This attitude lasted until Pearl Harbor Day, when people spontaneously gathered at the Center, seeking news and reassurance. However, even before the declaration of war, people in Greenbelt felt they should "do something." In August 1940 efforts began with a talent show to benefit the Red Cross. Several groups of refugee children spent two-week vacations in Greenbelt. The National Refugee Service proclaimed: "Greenbelt is the first community to attempt to do something for these children."[17]

By the time the war began, Greenbelters had several years' experience in organizing their lives through committees. In October 1941 town leaders set up a Defense Council—a "committee of committees"—to coordinate all war efforts and programs in the city. These included hosting Eleanor Roosevelt at a war bond rally on February 27, 1942. Greenbelt won the praise of Prince

American Legion Auxiliary wartime sewing center,
Greenbelt, 1940. Photo courtesy Greenbelt Homes, Inc.

George's County officials for the rally's success and also for cooperation
during blackout drills held throughout the winter of 1941–42. The American
Legion Post grew until it became the largest in the county.

In the early stages of the war, when Americans feared an invasion along the
East Coast, the American Legion set up an airplane observation post in
Greenbelt, one of twenty-four in the state. Airplane spotters took two-hour
shifts on the roof of the drugstore. The January 23, 1942, *Cooperator* noted:
"Many civilian posts, for want of personnel, have been taken over by the
Army, but our post is kept open every hour of the day and night, week in and
week out, by a volunteer force of 178, each of whom serves for two hours a
week." The utilization of nearby Schrom Airport as a training field for pilots
made the spotters' job more difficult.

Greenbelt residents threw themselves wholeheartedly into fund-raising
for war causes, salvaging scarce materials such as aluminum, and growing
victory gardens. They also coped with loss, as by late spring 1944 six Greenbelt men had died in service. According to their memories published in

*Looking Back,* women whose husbands were overseas thought that Greenbelt was a good location in which to endure the war. They could easily maintain the small housing units, and assistance to those in need quickly appeared. "I still remember and appreciate what all the older couples did for us wives, when our husbands were away," one commented. Another said, "After we had been here a short time and my husband had to go overseas, I found that many of my neighbors were single mothers with children. We had a wonderful time together. We helped each other out and would take turns taking the children in the mornings, so the other mothers could go shop or get away. So it worked out very well, and we had a social life within our own group."[18]

Greenbelt's busy adults, caught up in the war effort, still took time to oversee their children's education. Even though parents lacked actual control over the schools, they monitored all phases of school activity and prodded federal and county officials to action whenever they felt it necessary. This task became more difficult as more federal agencies became involved. The FWA appropriated money for schools in connection with defense housing, while Center School came under the aegis of the FPHA, now that it administered Greenbelt.

Due to the doubling of Greenbelt's population, more schoolrooms were urgently needed, but the federal agencies involved produced new buildings at a snail's pace. Both a new elementary school in the north end of town and a large addition to the high school had progressed only to the planning stage when the United States entered the war. In September 1942 both Center School and Greenbelt High School began operating on double shifts to alleviate crowding. By September 1943 Center School had 700 students and operated on a double shift for grades one through eight. The 175 kindergarten students were taught on a triple shift. At the high school the cafeteria and space originally used as offices by the FPHA were converted into classrooms. Bombarded with parent complaints, in August 1944 the FWA finally began construction of new school facilities. North End Elementary School had twelve classrooms. The addition to the high school consisted of four classrooms, a science laboratory, and a small assembly room; they were completed in May 1945.[19]

Demonstrating their willingness to take action regarding the schools, parents living in the new north end refused to allow their youngest children to walk to Center School, as, without interior walkways and pedestrian underpasses, the youngsters had no safe route to school. In response, the federal government and Prince George's County Board of Education announced, in September 1943, that they would remodel six houses in the north end to make a temporary school for kindergarten through second-graders. This opened in October with 200 students. Housing units proved extremely versatile: they were used for a co-op nursery school, a kindergarten, a health clinic, a hospital, a food store, a school, and a child care center.

While resident cooperation continued throughout the war in most areas of Greenbelt life, cooperation did not flourish in health care or religion. The Health Association continued through the war years, but had a tumultuous existence even by Greenbelt standards. It had begun operation in 1938 as a voluntary, nonprofit cooperative to provide medical care for its members. It even set up a small hospital in several row houses. By August 1941 members had divided into factions over the proper way to run the Health Association, and the rapid turnover of doctors caused by the war added to the difficulties. In January 1942, when the FSA refused to put any more money into Greenbelt's hospital, the only one in the county at the time, it closed. (Leland Memorial Hospital in Riverdale opened soon after.) The FPHA ruled that independent physicians could open offices in Greenbelt, and some did, competing with the Health Association.[20]

The cooperative ideal regarding religion also changed focus during the war years. Before the war, the goal had been an interfaith building to serve all groups in the community. When the number of residents suddenly doubled, churches coped as well as they could, but eventually individual denominations began to look toward erecting their own church buildings. A February 11, 1944, editorial in the *Cooperator* titled "Church Building Progress" reflected the new view: "All residents of Greenbelt will be encouraged on learning of the increased activity among Catholics here for erection of a church. In the early days of the town there were ambitious plans for a joint church building for all the major faiths. Now after six years we would welcome any kind of a structure which would solve the services space problem of even part of our worshippers." This change in attitude was unusual for the newspaper, as it regarded the promotion of Greenbelt's cooperative ideal as its primary mission. The newspaper almost always accurately mirrored the opinions of Greenbelters, so resident desire for an interfaith building must have decreased with time. Perhaps the town's increased size made the hope for one interdenominational building unrealistic, as a fairly large building would have been needed to house all of Greenbelt's worshippers.

A Greenbelt victory garden, 1942. Photo by Marjory Collins,
U.S. Office of War Information, Prints and Photographs
Division, Library of Congress

Although Greenbelt's unique plan and cooperative society continued to attract media attention, this attraction lessened somewhat during the war. Journal articles reflect the political orientation of the author or journal as much as the reality of life in Greenbelt. The *New Republic*, in an evaluation of the town after three years, focused on its advantages. But an article in *Nation's Business* concluded that Greenbelt "is an example of the waste and ineptitude that results when Government undertakes to do a job that should be done by private enterprise." The *Christian Science Monitor* and the *New York Times* both noted the success of the co-ops.[21]

From 1942 on, with most people's attention centered on the war, only the Washington and Baltimore newspapers continued to cover Greenbelt, generally focusing on local events, as they would for any other town. However, even these articles reveal differing editorial opinion regarding Greenbelt. Throughout the war years, the *Washington Post* described Greenbelt in positive terms. The following headlines are typical: "Greenbelt Goes Sports-Minded in a Big Way; Everybody Has Fun," "Town Fair Opens at Government's Model Community," "Greenbelt about to Pay Dividends," "Greenbelt Is Model Worth Copying, Britons Declare," and "Greenbelt Starts Eighth Year Sturdily."[22]

The *Washington Star* appeared fairly neutral but was sometimes negative in its commentary, and the *Washington Daily News* remained hostile to Greenbelt, typical headlines being "Santa Claus Is Leaving, Greenbelt's on the Block" and "Rents at Three Greenbelts Don't Even Pay Expenses." An interesting contrast can be seen in the way the *Post* and the *Star* reported on the same issue. During the war, the incomes of many federal wage earners in Greenbelt rose beyond the ceiling allowed by FSA guidelines. However, with the town already in turmoil, the agency decided to let the families remain so as to provide stability. The *Post* reported, "Greenbelt Holds Back on Evictions, Families Income Rise, but Housing Shortage Balks Plan to Move Prosperous." The *Star* headline was "F.S.A. Is Held Lax in Enforcing Rules at Greenbelt."[23]

While the original Greenbelt tenants were used to being the objects of humor, it was a new experience for wartime residents. All of the Washington papers still poked fun at Greenbelt's rules, although this occurred with less frequency than in the first few years. In May 1943 the FPHA added a new clause to tenants' leases: "The tenant agrees to notify the government of any change in the composition of his household. Additional members shall not be permitted occupancy of the premises except with the written permission of the government." Because of the housing shortage during the war, the city administration feared that families would take in relatives or friends in need of shelter. Officials hoped to prevent such actions by their notice, but they caused the following headlines: "New Greenbelt Leases Suggest Babies Must Have U.S. Sanction," "New Greenbelt Babies to Need U.S. Approval," "Stork's Wings Get Clipped at Greenbelt," "Stork Must Have Uncle Sam's Okay."[24]

Greenbelt's unique design continued to attract visitors even during the war. In August 1940 the city manager estimated that 5,000 sightseers continued to arrive each Sunday. This number dropped during wartime, although special groups continued to come, such as University of Pennsylvania social work students, University of Maryland economics students, and the Archbishop of York, in town to visit the National Cathedral. In November 1943 two British officials, the chief architect and the director of postwar planning in the British Ministry of Works, visited. The *Cooperator* always made residents aware of distinguished visitors, reinforcing the message that their town remained a unique and special place, worth traveling many miles to see.

Greenbelt's unique plan itself caused the greatest challenge to community cooperation during the war. Having constructed a town divided into two neighborhoods, federal officials in fact created a town divided into old and new, defense and nondefense homes. Whether or not town residents successfully forged an identity as one town or remained divided into two parts is a matter of some debate. Presumably the efforts of townspeople to unite to fight the enemy abroad, as well as the designation of all as Greenbelters by outsiders such as journalists, helped to draw the two parts of the town together, but opinion varied greatly on how successful it was.

Some residents of the original homes felt no resentment toward people in defense homes and accepted them as part of the town, with comments such as "I don't believe there was ever any great feeling, any great resentment about the newcomers coming in. . . . I never felt there was any antagonism between the two groups." "No one felt resentment about it. They just didn't pay much attention to it. Everybody was so busy." A newcomer, when asked if there was much contact between new and original tenants, responded, "Yes, I would say so. It depended on the interest we had in common. We visited back and forth, and of course there were all those various committees that we served on together."[25] Others felt a division keenly, especially it seems, the parents of children at Center School. Some mothers discouraged, or actually forbid, their children to play with children from the defense homes. Some adults felt that the defense homes downgraded the community. They were built as "temporary housing," one said, and "we felt that the people in the defense houses were just sort of camping there for the duration of the war." Another described the new homes as "practically slum dwellings."[26]

There is no question that all Greenbelters joined together at war's end to celebrate in a spontaneous, massive street party. Immediately after the announcement of the end of the war, the air raid siren blared, and "the mobile portion of the populace streamed down to the Center to make a joyful noise over the victory in the Pacific." Members of the band arrived and with the boy scouts led the crowd in forming a parade, which made its way through town. "Committee meetings and practically all other business afoot at the moment were canceled in favor of spur-of-the-moment parties, impromptu dancing in the streets, and dashing off to Washington to join in the general rejoicing there."[27]

Only a few days later, interested citizens gathered at a meeting to discuss the future of Greenbelt. Residents realized that the end of the war heralded the dawn of a new era and wished to have a voice in the shaping of their future. Greenbelt at war's end was a very different place from what it had been in 1940. With its size more than doubled, its constant turnover in population, the division between the new and old parts of town, it would have been extremely easy for the unique aspects of Greenbelt to disappear. That this rapid growth took place during a period of war must have exacerbated the effect of change. However, in October 1944, 313 of the original 885 families still lived in Greenbelt.[28] The mayor who served throughout most of the war, as well as many others serving in leadership positions, had lived in the town since its creation. Those in the defense homes had much in common and helped each other as necessary, just as the people in the original homes had. And the many activities that made up the war effort helped to join these two groups together.

The town's co-op businesses flourished, even while adjusting to rapid growth. Some cooperative efforts such as the Health Association produced discord, while others, such as the nursery school, proceeded quietly. The newspaper vigorously promoted town traditions, with the Labor Day festival resuming at war's end. Many social organizations remained active, as did a number of religious groups. Participation in club, church, and town events served to draw new and old residents together. Even though the new "temporary housing" lacked the finishing touches of the earlier homes, the FWA had constructed it according to the original plans, so it blended well into the community.

By war's end the people in Greenbelt had lived through a period of dramatic upheaval, but an underlying stability existed. The original physical plan remained intact, the co-ops retained their vigor, and residents continued to organize for every conceivable purpose in pursuit of their goals. The tenets of a cooperative life remained vital. With the prospect of peace, Greenbelters met to plan their future as, unbeknownst to them, officials of the federal government met to plan the sale of their town.

# 4

# The Government
# versus Greenbelt

DURING THE TEN-YEAR PERIOD following the war, Greenbelt
suffered repeated attacks by its creator, the federal government. These came
in several forms and from different government agencies, beginning with a
U.S. House of Representatives investigation into cooperatives. Following this,
the government secretly made plans to sell the town, which it accomplished
after the interruption of the Korean War, leaving residents struggling to form
their own cooperative housing. In addition to these traumas, several Greenbelt navy employees lost their jobs, fired as "security risks." The people of
Greenbelt, used to having their town branded as socialist or communist,
found themselves dragged into the widening net of McCarthy-style innu-
endo, accusation, plot, and rumor. Greenbelters, faced with harsh challenges
to their newly developed cooperative way of life, responded by uniting to
maintain their community.

Greenbelt residents focused their efforts on maintaining vital community
institutions such as the co-ops and the newspaper. However, achieving unity
to withstand attack from the outside required much effort as well as a willingness to compromise, which did not happen easily. Forced to become more
cohesive to ensure their continued existence as Greenbelters, they forged a
sense of identity, becoming more of a true community. Their experience

reflected Martin Buber's definition of the making of "a community of spirit" by beginning as "a community of tribulation."[1] How and why did Greenbelt residents, in a period of tribulation, emerge as a "community of spirit"?

As a harbinger of their decade of difficulties, in August 1947 the U.S. House of Representatives Small Business Committee undertook a study of cooperatives because it was uncomfortable with their increasing prominence. Committee chairman Walter Ploeser declared that cooperatives acted as unfair monopolies, which stifled competition and were thus "un-American." Chairman Ploeser claimed that "there has been an attempt on the part of some to impugn the purposes of the committee in its study of co-operatives. These hearings were not conceived to harass anyone."[2]

Greenbelt Consumer Services was the first cooperative in the country investigated by the committee. The first witness, Justice of the Peace Thomas Freeman, spoke for the American Legion and the Independent Trade Association of Prince George's County on the deleterious effects of Greenbelt co-ops. Next appeared Robert Black, the president of the Maryland Economic Council, who said, "If the Government continues its present policy of favoring co-ops, private enterprise will be forced to suspend or liquidate." The general manager, the office manager, and the treasurer of Greenbelt Consumer Services also testified. Other Greenbelt supporters traveled to the hearing in the co-op pantry bus, a vehicle ordinarily used to make shopping more convenient for North End residents by literally bringing the store to them. After two days of testimony the committee decided, two to one, that a monopolistic contract existed between the government and the cooperative and that private business should be allowed to enter the community. At the time, the government lease prohibited businesses of the same type as those run by the co-op from existing within a half mile of the business district.

Co-op supporters claimed that the city simply regulated the placement of business, as other cities do, usually by zoning ordinances. Representatives of the Federal Public Housing Authority (FPHA), the government entity responsible for Greenbelt, appeared before the committee, stating that they had no intention of changing their handling of Greenbelt due to the wishes of two members of the Small Business Committee. Thus this investigation changed nothing in the operation of Greenbelt co-ops. However, it served to demonstrate the unfavorable attitude of many in Congress toward Greenbelt.[3]

Even before the investigation by the Small Business Committee—in fact, as early as 1944—federal officials revealed their antipathy toward Greenbelt by resolving to sell the town as soon as possible. In 1944 and 1945 officials of the FPHA secretly discussed the upcoming disposal of the town with city councilmen. In late 1945, when the townspeople found out about the discussions through a leak to the *Cooperator,* they felt angry at being left in the dark not only by the government but also by their own elected representatives.[4]

On February 15, 1946, the *Cooperator* ran a lengthy editorial on Greenbelt's current situation. The last paragraph provides a succinct statement of the town's unique qualities, purposes, and goals for the future:

> The idea behind Greenbelt existed long before the men who planned our town were born, and will survive regardless of what happens to this particular band of Maryland landscape. The writers who derisively call it "Utopia" cannot be aware of how near they come to the truth; Greenbelt's greatness lies not only in the possibilities it has offered to all who have come to live in the community, but also in its actual accomplishments. Out of worn out tobacco fields and low income families, Greenbelt has built a town of tomorrow whose advantages are apparent even to its detractors, and a citizenry new in the realization that they have something here worth fighting for together. Greenbelt today is more than a Government housing project, more than a model of community planning, more than a headache for bureaucrats. Greenbelt is people who have had a chance at decent living, neighborliness, and cooperation, and who see no reason why, if the present arrangement can't continue to give them what they want, they should not be allowed to get it for themselves.

The crucial element in this statement is the desire for self-determination. The creators of Greenbelt within the Resettlement Administration and the Farm Security Administration (FSA) had encouraged the development of this attitude among the activist residents, and it continued to flourish.

The demise of the FSA during the war years demonstrates the change in attitude taking place within the federal government, which soon directly affected Greenbelt. The FSA staff encouraged collectivist, cooperative programs. By 1942 a strong opposition, consisting of commercial farmers, the Farm Bureau, and private processors and retailers forced to compete with co-ops, began to mobilize. In early 1943 U.S. Representative Harold Cooley of North Carolina headed a select committee to investigate the FSA. The committee found that the agency wasted money, was "communistic" and "un-American," and abolished it in 1946.

Although the management of Greenbelt had been transferred to the FPHA in 1942, the obliteration of the FSA clearly demonstrated a new federal attitude. As the historian Paul Conkin stated: "It meant that experimentation and broad attempts at institutional reform were ended."[5] Federal officials originally described Greenbelt as a great social experiment. By the end of the war, reflecting the new attitude, officials regarded Greenbelt as a collection of houses that the government no longer wished to own. However, it took the federal government almost a decade to manage the sale. The uncertainty engendered by this drawn-out process must have made the situation even more of an ordeal for residents than a quick decision, promptly administered, would have been.

Greenbelters found an ally in FPHA official Oliver C. Winston, who wished to expand the greenbelt towns to the size contemplated by the original planners or to sell them to institutions that would do this. He hoped the towns could be maintained as outstanding examples of community planning. As the government would clearly put no further funds into the greenbelt towns, Winston attempted to interest private foundations in the effort, to no avail. He was then lost to the Greenbelt cause in May 1947, when President Harry Truman reorganized all federal housing agencies, including the FPHA, into the Public Housing Administration (PHA), and Winston left the federal government.[6]

Interested observers, in particular Clarence Stein, took up where Winston left off, attempting to ensure that the sale of the towns occurred in such a way that their original design could be not only preserved but even extended. Stein lobbied officials of the PHA as well as key senators. Paul Douglas, chairman of the subcommittee involved, proved especially receptive to Stein's efforts.[7] The architect Catherine Bauer commented later that the "bill for the disposal of the Greenbelt towns to veterans' cooperatives through negotiated sale . . . was practically engineered by Clarence Stein with Senator Douglas."[8]

Public Law 65, signed by President Truman on May 19, 1949, allowed for the transfer of the greenbelt towns' real estate to private owners, with veterans' groups given preference. The law stipulated that current occupants should be allowed to join such groups. It also contained a provision allowing the PHA commissioner to transfer public buildings, streets, parks, playgrounds, and open land adjacent to the towns to nonfederal government agencies acting in the public interest.[9] Thus Public Law 65 provided the opportunity for the towns to be sold in a cohesive manner, which would maintain their original character. Evidence that the sale would proceed surfaced in "official notices" that appeared in local newspapers in June 1950, headed "To Veteran Groups Only." However, negotiations between the government and possible purchasers ceased for the two years of the Korean War.

In 1952 the time arrived for Greenbelt residents to show not only their willingness but also their organizational ability to purchase their homes as a group. Complicating the process was the fact that the community divided into two factions after the war. The divisions in Greenbelt in 1952 did not reflect the earlier partition into "old Greenbelt" and "defense homes." The fact that the two parts of town would be sold jointly presumably had a unifying effect. Instead, after the war, a new schism emerged—between veterans and conscientious objectors. The returning veterans had special influence, as demonstrated by the fact that the town would be sold to a veterans' group. A grateful country willingly provided special benefits for ex-servicemen, with most agreeing that they deserved them.

However, a number of pacifists, who had refused induction into the military due to their beliefs, resided in Greenbelt, drawn to the town by its cooperative nature and tolerant atmosphere. Some had been residents before the war; others came afterward. The town's co-op stores, unlike most employers, hired pacifists.[10] A number of former conscientious objectors were active in town affairs, and some held positions of leadership.

The conflict between veterans and pacifists emerged at a meeting of the Citizens Association on September 10, 1945. The purpose of the meeting was twofold: to discuss the future of the town and to provide a forum for fifteen candidates for town council. The meeting proceeded smoothly until James Flynn, an attorney for the Veterans Administration who did not reside in town, took the floor to question Walter Volckhausen, who was a city council candidate, an active member of the community, and a pacifist. Flynn asked Volckhausen if he would take the oath of allegiance if elected and if he had been investigated by the Federal Bureau of Investigation during the war. Volckhausen said he would take no oaths, that he had been investigated by the FBI, and that he had been cleared of any suspicion of being connected with communists. In spite of repeated requests, Flynn would not relinquish the floor and began to question Volckhausen about his religious beliefs. A supporter of Volckhausen, Irving Rothchild, called Flynn a "fascist," and in response Flynn swung at Rothchild, breaking his glasses and bloodying his nose. Others rallied around Flynn, the chairman could not restore order, and most of the audience simply left the room.[11]

The incident sparked an outpouring of letters by Greenbelters to the *Cooperator.* A number of letters to the editor supported Volckhausen, and others denounced the bullying tactics of his opposition. Reverend Robert

Kincheloe, former pastor of Greenbelt Community Church, provided the *Cooperator* with a letter he had written for Volckhausen to the Bureau of Selective Service. He summarized his statement, published in the September 14 *Cooperator*, this way:

> I respect the active and intelligent work of Walter Volckhausen in the Cooperative movement. It is his sincere conviction that through this movement there may be affected in the local, national and world communities a practical, down to earth approach to brotherhood, economic betterment for all, good will and peace. I know of many pious Christians, correct in their creedal recital and liturgical formalities, who have contributed practically nothing to the Master of Men defined as Religion. And I know a few men such as Volckhausen who have contributed much.

Someone else commented:

> It takes one kind of guts to go marching off to war. It takes another kind of guts to stick by your beliefs, regardless of consequences, knowing that you'll be misunderstood and hated by many. Without in the least wishing to detract from the spirit of the many who accepted military service, believing that in this way they could best serve mankind, the position of the true pacifist, which always seems ridiculous and illogical in time of war, should also be respected. In the last analysis, the peace of the world depends upon the ideals and actions of men of good will.

Other letters in the same vein reveal as much about the people of Greenbelt as they do about the character of Walter Volckhausen.

A postscript to the episode appeared in the next week's issue of the paper. The account of the meeting had described Flynn as a "Legionnaire." In a September 28 editorial entitled "Oops, Our Mistake," the paper admitted that "irate members of the Legion have been deluging the office with demands for a retraction, which we hereby make. Mr. Flynn is not a member of the Legion, and has never been." The editorial also commented on the praise the paper had received for the fairness of its article on the incident. It revealed that the man who ended the evening with smashed glasses and a bloodied nose, Irving Rothchild, was there in his capacity as *Cooperator* reporter and had written the article.

With people literally coming to blows, how could the citizens of Greenbelt ever hope to form one group to buy their town? A strong factor existed in their favor: the history of co-ops in Greenbelt. As the city manager explained: "Co-ops had existed from the beginning, and they were accepted. It was a logical extension from the other cooperatives in the city for a housing cooperative to be formed."[12] An early resident pointed out that most people involved in the effort knew each other and had worked together before, which helped considerably.

Residents even had previous experience in forming a housing cooperative. Just before the war a group of Greenbelters who wanted to stay in the town but who wished to have larger homes for their expanding families formed the Greenbelt Homeowner's Cooperative. They proceeded to clear land themselves in August 1941, having obtained a lease from the Farm Security Administration. Henry Klumb, a disciple of Frank Lloyd Wright, drew up plans for the group.[13] They selected a contractor and anticipated that twenty homes would be started in February 1942; however, U.S. entry into World War II halted their plans. Some members of this Greenbelt Homeowner's Cooperative remained in town and hoped to renew their efforts after the war. Their knowledge and experience undoubtedly proved useful to the two later groups interested in housing.

A group representing veterans and another representing town activists both announced their intent of forming a housing plan at war's end. In August 1945 the Greenbelt American Veterans Committee began consideration of veterans' housing needs. In November, at the monthly Citizens Association meeting—which was regularly attended by town activists—the Housing Committee recommended that the association "express itself as favoring operation of Greenbelt by a mutual ownership corporation consisting of residents and prospective residents."[14] Nine individuals elected by those present had the task of studying the possibilities for group ownership. They recommended the formation of the Greenbelt Mutual Housing Association, which was officially organized on July 15, 1946. This association, sponsored jointly by the Greenbelt chapter of the American Veterans Committee and the Citizens Association Housing Committee, brought together the two conflicting groups seeking co-op housing. The Greenbelt chapter of the American Veterans Committee represented the wishes of the returning World War II veterans, while the individuals most actively interested in developing co-ops, which included conscientious objectors, expressed their views through the Citizens Association Housing Committee.

In July 1946, at a meeting attended by 300 residents, Colonel Lawrence Westbrook, "nationally known expert on mutual housing," explained the benefits of mutual housing, in which the citizens of Greenbelt would form a corporation to buy the town and then lease the houses to member-residents.[15] In August more than 90 volunteers, working as block captains in their own courts, canvassed their neighbors to enlist members in the Greenbelt Mutual Housing Association. The August 9 *Cooperator* reported that 50–70 percent of those contacted signed up. Most who refused to join the association did so because they planned to move away. About 10 percent wished to stay but did not wish to join and, instead, formed a small group working to keep the federal government involved. By late September 640 tenants wished to participate; by March 1948 more than 1,120 families had signed up.

When it became clear with the passage of Public Law 65 that the federal government wished to sell Greenbelt to a veterans' group, the Greenbelt Mutual Housing Association metamorphosed into the Greenbelt Veterans Housing Corporation (GVHC), with 1,200 member families by March 1950. The board of directors hired a law firm familiar with public housing to handle relations with the PHA. Demonstrating their intention to remain faithful to the original plans and goals of Greenbelt, housing activists twice invited Hale Walker, one of the town's original planners, to speak. In August 1951 and again in October 1952 Hale Walker talked to more than 100 members of the GVHC, explaining his designs to bring the town to completion.

Walker's vision for Greenbelt featured additional areas or neighborhoods, replicating the original housing. His designs showed a small number of individual houses and duplexes, with most of the new housing being town houses facing courts, as in the original design. His plans emphasized two major areas of new housing, one north and one south of the lake, with the area south of the lake having a school and stores in the center, replicating original Greenbelt. Lots were set aside for churches, exactly where the current Jewish synagogue and the Methodist, Lutheran, Community, and Catholic churches are now located. A belt of green would surround the housing area, with allotment gardens to the east; a narrow park on both sides of Greenbelt road to the south; a park, the water treatment plant, and a cemetery to the west; and the Agricultural Research Center to the north. Land at the far western part of Greenbelt was zoned for industry.[16] A copy of this plan remained on the wall in a prominent location in the city manager's office for the next twenty years, demonstrating the continuing intention to remain faithful to the principles of the original plan.

The GVHC had a plan drawn for the city's future that followed similar principles.[17] Since only half of the plan remains, a direct comparison is difficult. However, the existing half contains a statement of "Principles of Land Development" with the subheading "to make Greenbelt a safe, convenient, economical and pleasant place to live." These principles, quoted below, remained key to the wishes of Greenbelt residents for the continued planning of their town:

General—All Areas:

1. Pedestrian walkways separate from through streets—especially along routes where children go to school or playground.

2. Park buffers between land uses—especially between residential and non-residential or heavy traffic arteries.

3. Roads differentiated by use. Through traffic streets to have minimum side street and driveway interference. Residential streets designed to prohibit through traffic.

4. Ample off-street parking in all areas. At least one off-street space per dwelling. At least three times as much shopping parking area as shopping floor area.

5. Preservation of Greenbelt principles wherever possible—especially shopping facilities fronting on landscaped pedestrian malls and living areas of homes turned towards park areas and view.

These general goals clearly emanate from those the federal planners used to create original Greenbelt. That the GVHC wished to continue them signals the satisfaction of the community with their use.

The GVHC plan also continued Perry's neighborhood unit with the following requirements:

Neighborhood Planning:

Each Neighborhood to have:

1. Boundaries defined by parks or major highways.

2. Population sufficient to require and support its own elementary school.

3. School, park and playground areas to meet all foreseeable requirements.

4. Internal off-street walkway system between school, playgrounds, and homes.

For developing the new areas of town the GVHC plan specified the following as guidelines for individual homes:

1. Off-street walkway and park system combined with storm drainage system.

2. Large blocks, with park areas generally within them and not bounded by streets.

3. Minimum permitted lot width at building line for detached houses 70 feet, desirable minimum lot 75 × 100 feet.

4. Existing natural features to be retained—specifically a minimum of 50% of trees and ground cover.

The GVHC plan added a section of recommended legal controls to be sure that development proceeded as Greenbelters wished:

1. Subdivisions to be developed only after land use pattern for neighborhood established and approved by Maryland–National Capital Park and Planning Commission and city of Greenbelt, with street and utility engineering completed.

2. All land to be covenanted before sale to insure continued use in accord with development program.

(In an omission that proved disastrous within a decade, future members of the GVHC neglected to follow these recommendations regarding legal control of the land.) The adherence of both Hale Walker's plans and those of the GVHC to the original design and goals of Greenbelt suggests that Greenbelt residents could clearly articulate their wishes to continue in accordance with the original design. When the citizens of Greenbelt referred to their goal of keeping Greenbelt a planned community, they had in mind the ideas suggested by the Hale Walker and GVHC plans.

In February 1952 more than 500 members attended the annual meeting of GVHC to elect its board of directors. The GVHC board signed a purchase agreement with the government (after rejecting its loyalty oath clause) on December 30, 1952, purchasing 1,580 dwellings and 709 acres of vacant land suitable for housing. The government sold the 310 apartment units separately, but in 1953 the cooperative purchased 60 of them, along with 52 garages. The shopping center went to the highest bidder in 1954, while 1,362 acres of the greenbelt were transferred to the Department of the Interior, becoming Greenbelt National Park. Private buyers purchased several smaller parcels of land. The federal government realized $8,973,767 from the sale of the town.[18]

GVHC's purchase was made possible by the financial assistance of the People's Development Company, a subsidiary of Farm Bureau Insurance Companies. This cooperative insurance company provided the $150,000 needed to make the down payments on unsold houses, as well as another $85,000 for the down payment on the undeveloped land, which was regarded as a good investment. Banks and savings and loan institutions would not deal with a housing cooperative, so the assistance of the People's Development Company was crucial.[19] A December 31, 1952, editorial in the *Cooperator* reflected on the GVHC's accomplishment: "Their goal has been a Greenbelt owned by the residents and not exploited for profit; a Greenbelt developed as the original planners meant it to be, a demonstration of cooperative living at its best." Greenbelt residents must have felt great satisfaction at achieving their goal after so many years of effort.

However, their struggles had not ended. Disagreements as to the running of the housing co-op occurred frequently in 1953 and 1954. As was usual, differences of opinion were aired in letters to the editor of the *Cooperator*. On June 18, 1953, the *Cooperator* printed an editorial entitled "The Orgy Is Over," which began: "For several weeks, this newspaper has printed letters to the editor pertaining to squabbles among the members of the GVHC Board of Directors. We have printed every letter received, and we have shortened none of them. Week by week, these letters have become more vicious, more venomous, more personal. . . . We approve the public discussion of differences of opinion when based on principle, but not on personal animosity." The editor declared that no more letters would be printed that "cast reflections on an individual's character or in any way question his integrity." But as early as March 1954 the paper again devoted whole pages to letters to the editor regarding the operation of the GVHC. Although members had united to jointly purchase their homes, once past that milestone they continued to experience difficulties working as a group. The issues of whether or not to allow pets and the setting of co-op fees caused the most disagreement. (After much waffling and several votes, pets were allowed.)

A considerable shuffling of population took place after the sale, with families leaving for several reasons. Those who had moved to Greenbelt due to their defense-related work in Washington had to leave at the end of the war, while others simply wanted bigger homes. GVHC canvassers had discovered that some disliked cooperatives on principle. One co-op activist remembers receiving this response to an appeal to join the co-op: "Hell, no,

Co-op grocery delivery bus, Greenbelt, August 1947 (group
having lunch has been attending congressional hearing on
Greenbelt Consumer Services). Photo by Nick Pergola

that's socialism!"[20] Nonpurchasers of GVHC homes, about 600 of a total of
1,600, were required to leave by August 31, 1953.

Hoping to fill these 600 units, the GVHC advertised in local papers, such
as this one in the *Washington Post–Times-Herald:* "Greenbelt, a Beautiful
Suburban Planned Community with unusual recreational facilities, swimming pool, playgrounds, picnic areas, complete shopping center, schools
and churches. Brick Homes $6,000 to $9,500." "Prospective Buyers Flock to
Greenbelt," the *Post* declared, reporting that "several thousand house-hungry
Washington area residents thronged Greenbelt, Maryland, yesterday, as
about 600 bargain-priced homes in the community were put on sale."[21] Just
as when Greenbelt was originally built, potential residents converged on the
town, attracted by its affordability relative to other housing in the Washington, D.C., area.[22]

The Community Coordinating Committee, consisting of representatives
of twenty community organizations, "was formed to acquaint new residents

with the community and to undertake projects which would not be possible for the separate organizations to tackle."[23] The committee began with the distribution of information about Greenbelt and its organizations. Its major effort was a citywide party held on May 21, 1953. The committee mailed letters of invitation to all newcomers, but to make sure no one was missed, "old residents are asked to corral all the new residents in their courts and escort them to the meeting." Games and square dancing provided the entertainment, with lots of food, as well as information on community activities.[24] In its May 28 follow-up article, the *Cooperator* did not say how many participated but noted that all 1,200 cookies as well as other refreshments disappeared. Greenbelters thus attempted to make the newcomers feel a part of the community as rapidly as possible.

The sale of Greenbelt not only led to an infusion of new residents but also brought two previously separate groups, those in the original community and those in the defense homes, into one unit, the GVHC. This association seemed to end the previous differentiation between the two groups. The sale also forced other opposing groups to work together for a common cause: veterans of the war and active supporters of cooperatives, including pacifists. Despite the differing natures of the people involved and the endless debate over exactly how the co-op should be run, a core group of activists emerged who were concerned with and willing to devote themselves to the long-term future of Greenbelt.

In May 1957 members of GVHC changed its name to Greenbelt Homes, Incorporated. The original name had been necessary under the conditions of sale by the federal government, in which veterans' groups received preferential treatment. However, by 1957 the emphasis on veterans had turned into a liability, since any adult could become a co-op member. In 1958 Greenbelt Homes included 1,500 member families. Interest remained high in the running of the organization; in 1959 fifteen individuals ran for the nine places on the board of directors. Greenbelt Homes advertisements in local newspapers proclaimed, "Living is Great in Greenbelt." The co-op credit union continued to prosper, and a new cooperative, Twin Pines Savings and Loan, began operation in 1957 to help people finance their Greenbelt Homes houses.

Throughout the period of the long, drawn-out sale of their town and the formation of the housing cooperative, Greenbelters bore the brunt of another attack on their struggling community as they were pulled into the morass of McCarthyism. It was probably inevitable that Greenbelt was swept up by the tentacles of the 1950s witch-hunt, as the town had been labeled communist even before it was built. Throughout Greenbelt's short history, residents became accustomed to their cooperative way of life being branded as socialist or communist.

In a prescient article entitled "Still Witch-hunting," the October 19, 1939, *Cooperator* editor commented on the renewal of funds to the Dies Committee, even though

> witnesses before the committee were discredited in numerous cases, evidence used was often seized by dubious means—by theft, bribe and assault—hearings were sometimes secret, and even the open sessions run in such a haphazard and biased manner that leading lawyers and Congressional figures commented openly on the fraud, the indictments made were never presented to the accused, nor were the victims of these broad accusations ever allowed to appear in their own defense.

The editorial concluded, "All but the super-patriots know that any dangers to our country can be ferreted out and prevented by the thoroughly capable Federal legal and penal forces at our disposal. When partisan politicians set themselves up as police force, prosecutors, judges, and jury we need to take care that what we laughed at yesterday does not become a menace tomorrow."

On May 2, 1940, the paper, in an article on the Nazi takeover of Norway, concluded: "FBI investigation of the KKK, certain fascist army leaders, certain reactionary Congressmen, and a certain 'radio priest' might be at least as fruitful as recent rough-shod investigations of liberal, progressive, and radical organizations and individuals." Fears of a "communist threat" flared anew in the postwar world. A *Cooperator* editorial on June 10, 1948, titled "Confusion and Fear," began, "Not since the red scare that followed the first world war has such widespread fear and confusion surrounded the communist problem." In 1950 the August 10 issue reprinted in full President Truman's address to Congress, which concluded, "We must not be swept away by a wave of hysteria."

The atmosphere of suspicion and distrust fueled by fears of communism expressed itself in several ways in Greenbelt. The first to occur involved two incidents of censorship in 1951. An Italian film called *Bitter Rice,* slated to be shown by the Greenbelt Consumer Service's movie theater, was withdrawn after protests, mainly by local Catholics. The film received a "C" rating by the Legion of Decency, thus forbidding Catholics to view it. The withdrawal prompted cries of censorship by many others, and the theater showed the film the following week. In December 1951 a city council member pointed out that the city library regularly received copies of the *USSR Bulletin*, a propaganda newsletter sent to every library in the country. He moved that the city withdraw from the mailing list. Mayor Lastner felt that "we would be doing what they are doing" by denying free access to printed material. In a vote of three to two, the council decided to continue to receive the publication.[25]

In September 1952 Walter Volckhausen questioned the loyalty clause inserted into the GVHC by-laws at the insistence of the PHA. In a lengthy letter, printed in full in the September 18 *Cooperator,* he summarized his thoughts this way: "The loyalty program is a sinister invasion of the freedom of thought for which our Nation has been noted; its victims are far more often those of independent judgment than those of malicious intent. It makes neighbor suspicious of neighbor, and it hampers intelligent consideration of vital local, national and world problems." Other letters printed in following weeks agreed with Volckhausen's analysis. He must have reflected a fairly general feeling, because when the GVHC signed the agreement with the PHA, the loyalty clause had been removed at GVHC's insistence.[26]

These admittedly minor incidents accurately mirrored the atmosphere of the time. From his speech in Wheeling, West Virginia, on February 9, 1950, to the Army-McCarthy Senate hearings in April through June, 1954, Joseph McCarthy created a national atmosphere of a witch-hunt as he diligently searched for communists.[27] According to the historian Richard Fried, a key element of what became McCarthyism occurred in 1947 with "the decreeing of the federal employee loyalty program. This act helped fix the assumptions, language, and methods that fueled the assault on American liberalism mounted by McCarthy and other anti-Communist politicians. . . . Truman did not invent so much as codify, institutionalize, broaden and tighten FDR's jury-rigged wartime program."[28]

According to the historian George Callcott, after 1944 a number of groups busily sought evidence of communism in Maryland. They included conservatives, Catholics, military groups, and businessmen.[29] This appeared true in Greenbelt, where the Catholics and the American Legion concerned themselves with the threat of communism. Greenbelters made a likely target, not only because of their town's cooperative nature but also because of the large numbers of both federal and military employees in town. One such employee, Abraham Chasanow of the Navy Hydrographic Office in Suitland, Maryland, upon being called to the personnel office on July 29, 1953, received three letters. The first informed him of his status as a suspected security risk.

The second said that "he would henceforth be denied access to classified material. The third said he was suspended from duty without pay." He responded, "If it's taken the Navy 23 years to decide I'm a security risk, God help the Navy."[30]

An examination of his work record made him seem an unlikely suspect. Chasanow worked in the Hydrographic Office, which handled the navy's charts. During his twenty-three-year employment, he made his way to the top of the division, with 63 employees under him. He won two Meritorious Civilian Service awards during these years and was 1 of only 5 employees out of 1,500 to receive a rating of outstanding in 1951. Several times his suggestions on ways to tighten security had been adopted. During World War II he took steps to stop the sale of navy charts through private channels. He had recently ordered his office to keep all material hidden from a visiting Russian mission.

On a personal basis, he also appeared an unlikely type for a security risk. Along with his wife, Helen, and their four children, he actively involved himself in the community. Chasanow had been the first male Parent-Teacher's Association president in Greenbelt, he had headed both the Citizens Association and the Lions Club. Helen Chasanow had led Greenbelt's Community Chest and Red Cross drives. A few weeks before his suspension, Chasanow's daughter Phyllis, a previous "outstanding Girl Scout," won a Girl Scout essay contest, which she concluded with the line, "There is a glow in my heart because I am an American." Chasanow guided the fight to retain the loyalty oath in the GVHC contract, and he was even a Republican in largely Democratic Greenbelt.[31]

The government charged two other men in addition to Chasanow: Isidore Parker and Mike Saltzman. They rode to work with Chasanow and their cases appeared to be guilt by association, with Chasanow being the main target. Saltzman did not fight the charges, but left Greenbelt for California. Parker fought to retain his job, as did Chasanow, but his case never became well known, as he did not choose to utilize the media as Chasanow finally did.

Why were these men selected as targets? They had several things in common: all lived in Greenbelt, all were Jewish, and all worked for the navy. During the McCarthy era it appeared that the Office of Naval Intelligence (ONI) was especially desirous of finding communist sympathizers. Another Greenbelt resident, Bruce Bowman, expected to be "called out" at any time, because of his activities in town, but he never was. He worked for the Army

Corps of Engineers, which apparently did not have the same zeal to locate communists as the ONI.[32] At least twenty other Greenbelt residents mentioned in the charges against Chasanow—many of whom held high-ranking jobs with the army, air force, and government agencies—did not undergo investigation.

However, Chasanow and his lawyer believed there was a more compelling reason that he in particular came under attack. During the turmoil of the sale of the town, a minority group formed to protest the sale to the cooperative, thinking that the federal government could be convinced to maintain the town and that they could thereby keep their low, subsidized rents. Chasanow served as attorney for the GVHC, and after the co-op successfully purchased the homes, his name appeared on the letter sent to all residents informing them they would have to buy their homes or leave. Possibly those forced to leave town focused their resentment on Chasanow. It seemed credible that some of these men reported Chasanow to the navy.[33]

Greenbelters understood that not just a few individuals, but rather the entire town, was under attack, as demonstrated when the Naval Board of Inquiry discussed Greenbelt in its findings:

> From its inception Greenbelt has been a subject of controversy. From without it has been eyed suspiciously . . . as a "queer" experiment. . . . The extent of the cooperative undertaking was viewed askance by many and associated with something apart from private enterprise. The result has been that rumor and gossip has given Greenbelt a "radical" or "leftist" reputation. . . . There can be little doubt that there have been and still are persons living in Greenbelt that have radical tendencies.[34]

Many residents of Greenbelt responded to the attack on Chasanow and Parker as an attack on them all. Morally they felt such a stand to be necessary and correct, but they also realized that almost any of them could be next. The majority of the community appeared willing to support Chasanow publicly.

On the day Chasanow received his job suspension, he came home and pulled down all the shades at his windows. He said his first instinct was to hide in the dark. He "felt above all a pervading fear, a sense of isolation. At first you don't want to talk on the telephone. You don't want to say hello on the street. It's an awful feeling."[35] His feelings reflected an atmosphere of distrust that hovered above Greenbelt like a poisonous cloud. A GVHC activist recalled that some people would not speak to him on the street because he was a leader of that "radical" organization.[36] City manager Charles McDonald received criticism for involving himself in a security case while holding nonpartisan public office. Mayor Frank Lastner, on his way to confession at St. Hugh's Catholic Church, heard a fellow parishioner yell at him, "There goes another communist." The police chief George Panagoulis, at a get-together with federal law enforcement officers, was greeted with, "Nice to see you, comrade." Anthony Madden, a Greenbelt insurance agent and a Catholic, was told that his support of Chasanow would hurt his business.[37]

When Chasanow returned from work on July 29, in a state of shock, he felt the need for guidance. As his rabbi was not available, he sought out the pastor of the Greenbelt Community Church, Eric Braund. Chasanow was inclined to fight the charges, and Braund supported him in this. Fellow attorneys suggested Chasanow seek out Joseph A. Fanelli, former special assistant to the U.S. attorney general. Fanelli was an aggressive, Harvard-trained lawyer who had successfully defended several persons caught in the net of the Truman loyalty program. In an interview at the time, he commented that he enjoyed interesting cases and lowered his fees for them. "I let the dull cases pay for them."[38] Fanelli warned Chasanow that he would need affidavits as to his character and that, in the pervasive climate of suspicion, they could prove hard to get. Despite these fears, many of Chasanow's co-workers stood by him; a group came to his home bearing ice cream, cake, and candy. Twenty-four present and former Hydrographic Department workers swore affidavits for him, and more would have if asked. Chasanow collected an additional seventy-three affidavits from friends, most of them in Greenbelt.[39] While Fanelli prepared Chasanow's case and collected affidavits, the ONI conducted an investigation in Greenbelt that left several residents highly indignant. Chief Panagoulis, a graduate of the FBI school, remarked after their visit: "They had been all over Greenbelt. Their roamings, themselves, originated rumors. On the basis of my training and experience in investigative methods, I personally considered those two fellows to be rank amateurs in the field of security investigations." McDonald, the city manager, described his interview with the agents: "In their conversation they threw out various names, endeavoring to get me to verify that I had knowledge of Mr. Chasanow's close association with these people. I had no such knowledge. I explained thoroughly to them that Mr. Chasanow was an active, interested citizen. . . . The agents left somewhat disturbed that I had not given them the answers they had wanted." The reaction of Mrs. Winfield McCamy, the city clerk: "They kept trying to put words in my mouth about a radical group at Greenbelt."[40]

Chasanow's case went before the first hearing board summoned by the navy under the new Eisenhower security system. It lasted for three days, beginning on September 21, 1953. The ONI brought eight charges against Chasanow, most of them alleging that he associated with various persons thought to be communists. The most serious charge was that "several reliable informants have described you as a leader and very active in a radical group in Greenbelt." Thus a large part of the hearing involved a discussion of the Greenbelt housing dispute, management of the cooperative stores, and whether there was something "queer" about the *Cooperator*. Several witnesses from Greenbelt, such as Lastner, McCamy, and Madden, had their testimony presented by ONI agents in what they perceived as a distorted fashion, and they insisted on rebutting the "evidence" with their own.[41]

On October 9 the hearing board released its results, finding Chasanow's contacts with alleged communists to be brief and superficial. On the charge regarding radical groups in Greenbelt they found that "the weight of the evidence points to the nonexistence of 'radical' or 'left-wing' groups at Greenbelt. . . . In the case of Mr. Chasanow . . . the testimony showed that he was, if anything, a moderating, constructive and conservative influence." In sum, the board found that Chasanow's employment was clearly consistent with the national security.[42] The secretary of the navy received the results, and Chasanow expected to be promptly back at work, but he remained suspended without pay, month after month.

During this time of waiting the Chasanow family continued to receive the support of the community. By January 1954, in response to his deteriorating financial situation, Chasanow decided to open a law and real estate office in Greenbelt. The city council voted to lease him the old police station at a nominal fee. He obtained a bank loan with the help of a fellow townsman, the city contributed used furnishings, and he was soon in business. The Lions Club board of directors gave Chasanow a personal vote of confidence, and the GVHC officially endorsed him. Helen Chasanow offered to resign as vice president of B'nai B'rith to save the members embarrassment, but they refused her offer. The Women's Club insisted that the Chasanows attend their dances. Mayor Lastner explained: "They tried to hold back, but we wanted them with us. Abe was fighting for every decent person in town."[43]

On April 8, 1954, Chasanow finally received the call to report in for the official result of the hearing and to learn his job status. Since he had been exonerated by the hearing board, he was shocked to learn that he had been fired by the secretary of the navy. His shock soon turned to outrage, and, encouraged by Fanelli, he decided to "go public" with his story. On April 15 he held a news conference at Fanelli's Washington office, attended by reporters for all of Washington's major papers. On its front page, the *Washington Daily News* headlined "Man Here Fired as Risk by Navy after 23 Years Appeals to Public." The article by Anthony Lewis began, "A Navy civilian employee dismissed as a security risk despite a hearing board's strong finding that he should be cleared today petitioned for restoration of his job and took the case to the public."

To show their staunch support of Chasanow, six community leaders from Greenbelt attended the press conference: the city manager, Charles McDonald, the city clerk, Mrs. McCamy, the Reverend Eric Braund, the police chief, George Panagoulis, the mayor, Frank Lastner, and Morris Sandhaus, a former Greenbelt rabbi. Several of them told reporters about their experience with the navy investigators, who had questioned them about Chasanow and then had inaccurately reported their statements to the navy.

At the press conference, Fanelli stressed that the decision to take the case to the public had been made reluctantly because of possible harm to the Chasanow family, but they had concluded that it was necessary. Just the previous week President Dwight Eisenhower had stated that public opinion could be counted on to correct any violence done to the standards of fair play. Fanelli summed up: "We feel the Chasanow case is significant in part because it involves an unimportant man, and a man who by the loosest conceivable standard cannot be considered a risk to security. If Mr. Chasanow can be dismissed, then no American is safe."[44]

The decision to seek publicity paid off handsomely. Even before a series about Chasanow in the *Daily News* finished, Joseph Fanelli heard from J. H. Smith, assistant secretary of the navy, offering an informal hearing on Chasanow's request for consideration of his dismissal.[45] This was held in May, and on June 4 Chasanow received a letter from Rear Admiral Holderness disclosing that two charges had been dropped and that a special board would rehear the case. One of the dropped charges accused Chasanow of being a leader of, and very active in, a radical group in Greenbelt. On September 1, fourteen months after his suspension, the navy admitted it had made a mistake in firing Abraham Chasanow and restored him to his former position with full back pay. Assistant Secretary Smith said a "grave injustice" had been done to Chasanow and that a review of procedures in the navy's security program would be undertaken. Chasanow expressed his joy and relief: "I feel like I had

Congressional hearing on cooperatives, Washington, D.C.,
August 1947 (Dillon S. Myer, head of the U.S. Public
Housing Administration, *far left*, being questioned by
Representative Walter C. Ploeser, chairman). Photo by
Nick Pergola

awakened this morning from a bad dream and found the birds singing, the
flowers blooming and the sun shining."[46] On July 26, 1955, Abe Chasanow
decided to retire from his navy job to devote full time to his legal and real
estate work in Greenbelt. Fifty fellow workers gave him a goodbye luncheon,
from which higher officialdom was noticeably absent.[47]

To complete the story, Isidore Parker, the man charged along with Chasanow, also received his reinstatement, after which he retired and moved out of
Greenbelt. Parker worked as a cartoonist, and his art, which had been featured for years in the *Cooperator,* brought him celebrity in the community.
His cartoons appeared occasionally in the place of Herblock in the *Post* from
1964 to 1968.[48]

Why was Abraham Chasanow able to effectively and publicly clear his
name when many other, equally innocent individuals during this period,
could not? Several factors contributed to his success. The first and most
important was the Greenbelt community reaction. McCarthyism worked by
"shunning," making an individual a community outcast. The town of Green-

belt refused on principle to allow this to happen, even though Chasanow tried to isolate himself in order not to bring his troubles to others. Another factor was the timing. In April 1954, the date of Chasanow's firing, the Army-McCarthy hearings had just begun, and McCarthy's influence was on the wane. When Chasanow made his public appeal, the papers welcomed his story in a way they probably would not have earlier. The actions of the lawyer Joseph Fanelli and the reporter Anthony Lewis, who wrote a six-part series on Chasanow's experience, also significantly affected the outcome. The publicity generated by the articles evidently forced the ONI to evaluate its course, as it announced a change in position immediately after the series began appearing.

Greenbelt's experience with McCarthyism changed the lives of a number of individuals. Town residents read about themselves in the May 17, 1955, issue of *Look* magazine, which featured a lengthy tale of Greenbelt. The lead was, "This man was nearly destroyed by lies. . . . His neighbors saved him. . . . Here's a heart-warming true story that will restore your faith in the human race." The writer, Joseph Blank, emphasized the "goodness" of those in town. Anthony Lewis received the Pulitzer Prize for his series of articles in the *Daily News*. Twentieth-Century Fox purchased the screen rights, turning them into the movie, *Three Brave Men*. Ernest Borgnine portrayed Chasanow, with Ray Milland as Fanelli. Philip Dunn, responsible for *The Robe* and *How Green Was My Valley*, directed. Abe and Helen Chasanow spent ten days on the set in Hollywood, observing the filming. They returned with autographed photographs of Elvis Presley for their daughters.[49]

One last facet of the Chasanow case needs to be discussed because of what it reveals regarding the manner in which Greenbelters put cooperative ideology before personal preference. From all the support Chasanow received in the town, it might seem that he was a popular resident, but this was not the case; many individuals in town did not like him. He had taken unpopular stands, such as opposition to the removal of the loyalty oath in the sale of homes to the GVHC. His realty business, the only one in town, was viewed as directly competing with the GVHC. As the October 28, 1954, *Cooperator* commented: "The situation of the rivalry between the GVHC sales staff and the Greenbelt Realty Co. appears to be more of a serious problem than we had realized. There is definitely an undercurrent of antagonism there, although we're not prepared to say at this time where the main source is." Several individuals described Chasanow as "not a very pleasant person." Thus although many in town willingly supported Chasanow in his fight as a matter of principle, a number of residents did not get along with him or like him personally. The way people felt about him may have been an additional reason for his selection as a target; his unknown accusers may have thought he would have few defenders. After Chasanow's reinstatement, along with the tremendous feeling of relief that the town, as well as Chasanow, had been cleared, could be heard the remark, "Well, now I can go back to hating Abe Chasanow."[50]

Such unity of feeling that did exist occurred because of a common perception that the attack came not against one or two people but against the entire town. Thus support for Chasanow was based on the fact that he was a member of the community, not because of any personal characteristics. Residents of Greenbelt had become used to thinking of themselves as part of an interlocking whole. Most interacted continuously through their memberships in town organizations. Their Greenbelt ethic of cooperation demanded that they unite to fight a common enemy from outside.

The continuation of town traditions and group activity in the many co-ops, clubs, churches, and athletic organizations throughout these years of struggle helped cement personal relationships and encouraged interaction among all members of the community. This interaction and activity also played a crucial role in maintaining Greenbelt's identity throughout the period after the war. While the federal government now regarded Greenbelt as an unwanted burden, and the fear engendered by the McCarthy era brought criticism and suspicion to townspeople, Greenbelters continued as usual, relying on themselves to maintain their community.

The various cooperative businesses that composed Greenbelt Consumer Services experienced a period of great success after the war. In October 1945 the GCS paid off its loan to the Consumer Distribution Corporation, the Filene group that had provided seed money for the co-ops. After rapid expansion, by January 1949 the GCS was the fourth largest urban co-op in the United States in terms of sales volume.[51] In 1951 the GCS demonstrated continued growth by opening a supermarket and drugstore in Takoma Park. In 1953 the co-op joined in a venture with Rochdale co-ops to expand into northwest Washington. In 1954 "America's largest consumer cooperative, the Wheaton Co-op General Store, Service Station and Shopping Center," opened, another expansion of the GCS.[52] Thus by the end of 1954 Greenbelt Consumer Services had completed its metamorphosis into big business.

The evolution of the GCS into a typical business corporation and its subsequent loss of ties with Greenbelt set off a major feud with the other most influential co-op in town, the newspaper. The paper always covered the activities of the GCS in great detail and received most of its advertising from the GCS. This had earlier led to criticism that the paper was a "house organ" for the GCS, representing its interests rather than those of the townspeople. On August 22, 1941, an editor responded to such criticism this way:

> The charge that we are subsidized or controlled by this person or that group has come up so often of late that, for the benefit of those old-timers who may have forgotten and those newcomers who never knew, we should like to present our situation again. The *Cooperator* is a community paper, published by residents of the community without pay and given to the community without charge. Our only source of income is from our advertisers, the largest of them being our own Consumer Services. Inasmuch as you are the owners of Consumer Services, and the patrons too, the largest part of our income comes indirectly from you. This is almost the ideal situation.

On December 1, 1944, another *Cooperator* editorial appeared: "We are sometimes told that we treat the affairs of Greenbelt Consumers Services in detail . . . disproportionate to its importance. . . . We maintain that, as GCS is the largest single organization in town which touches the lives of all residents, it is pre-eminently newsworthy. We will continue, therefore, to the best of our ability, to keep readers posted on the developments and accomplishments of GCS." Thus the paper pursued the policy it felt to be right but was forced to acknowledge resident opinion.

Upon observing the GCS trend toward big business, on October 26, 1950, the watchdogs at the newspaper asked, "When does a co-op stop being a co-op?" This editorial criticized members for not paying attention to co-op activities and management for "forgetting" to consult members on major decisions. The paper published letters critical of the GCS; this one, dated February 22, 1951, is typical: "Last week GCS gave its membership only two days' notice of a special board meeting to consider the important question of whether GCS should immediately undertake additional expansion in the Takoma Park area." The writer objected to this expansion and explained: "The short notice and dire urgency for a decision is a sham to steam-roller the expansion through without giving the membership an opportunity to express themselves on the subject. . . . The membership, if given the opportunity to express themselves, would oppose expansion outside of Greenbelt because they need and want additional services and facilities in Greenbelt."

On April 19, 1951 the paper ran an editorial claiming that most people in town, except for the board and management of the GCS, agreed with their criticisms of the co-op. However, they also stated that this development did not please the paper staff, as "the truth is that we are for GCS, just as we are for cooperatives, and it is only because we are that we continue to try to improve it." In May 1951 the GCS board and the *Cooperator* staff met to try to improve relations, but the effort proved fruitless, and the GCS ceased to advertise in the paper.

The president of the GCS wrote a letter to the *Cooperator* defending the board's position. The letter was published on October 16, 1952: "Continual expansion is a basic principle of cooperatives, and of any business that wants to remain healthy. To provide ever better service, to maintain pace with expanding competitive businesses, to keep our most able employees, to make our cooperative more important to us and to others, we must move forward." The GCS board contended that expansion not only served others but also improved the quality of service to people in Greenbelt. Whatever the effect on service in Greenbelt, there is no doubt that the GCS grew beyond the control of its Greenbelt originators.

This loss of control and the subsequent weakened ties with the town earned the continuing disapproval of the paper staff. However, the adherence of the newspaper to its principled criticism of the GCS caused it to lose almost all advertising revenue, leading to financial difficulties. An editorial on May 17, 1951, stated: "Rumors have been going around Greenbelt this past week to the effect that the *Cooperator* is in difficulty. This is to assure our readers that the rumors are perfectly true." An editorial asked for advice from the community. Although Greenbelters differed regarding the paper's editorial policies and roles, unanimous agreement surfaced on the need to continue the paper. Buoyed by this acclamation, the paper carried on in spite of financial difficulties. In June 1953, delivery, which had been free, began on a subscription basis. In July 1954 the staff polled readers about a possible name change for the paper. They felt that the name *Cooperator* emphasized ties to the GCS, from which they now wanted to distance themselves. They received very little community response but decided to rename the paper the *Greenbelt News Review*.

In January 1955 the staff held a special meeting to discuss the future of the paper, which had a continuing shortage of both advertising revenue and volunteers. They decided to hold a communitywide meeting, asking all organizations in Greenbelt interested in the continuation of the paper to send a representative. After this meeting a February 3, 1955, editorial commented, "This newspaper reminds us of Sarah Bernhardt, the great actress who made about a dozen 'farewell' tours before she finally decided to throw in the towel. The *News Review* has been on its death-bed for so long that the springs are beginning to creak slightly, and the doctor is surreptitiously looking at his watch and thinking of other patients who need him more." In spite of this gloomy note, in February the paper began free citywide distribution once again, as community organizations provided their promised support by placing advertisements in the paper. In September 1955 an editorial again stated the need for more advertising and more volunteers for the paper to go on. For four years, without GCS advertisements, the continuation of the paper proved a weekly struggle.

Despite financial problems, throughout the postwar period the newspaper maintained its role as informant and dispenser of community information. It also carried out its mission of regularly reminding readers of Greenbelt's history. After the war, the paper discontinued the columns "One Year Ago" and "Five Years Ago," but in 1948 a new column, "Ten Years Ago," made its first appearance. In 1951 the paper published a series of articles on its own history.

The newspaper continued to play a major role as community conscience and prompter, stimulating interest in directions it thought proper, publicizing actions that needed to be taken, and invoking mass response when necessary. A series of articles about the Greenbelt Citizens Association provides a good example. A March 3, 1949, editorial asked: "What has happened to the healthy, sturdy, squalling infant that was the Greenbelt Citizens Association at its birth in 1937 and for the first few years of its life? Those who recall that lusty young voice must have difficulty in recognizing the whispers of the prematurely senile organization that is GCA in its twelfth year." On July 7, another editorial encouraged everyone to attend the meeting and election of the Citizens Association. The July 14 issue gave prominence to an article about the election and meeting, successfully held with sixty people in attendance.

On August 9, 1951, an editorial expressed concern at an apparent lack of interest in the coming city council election: "Voting is a sacred privilege. From the topmost level to the local scene, it is the duty of citizens to exercise their franchise. The deadline for registration is August 19, only 10 days away!" Two weeks later, on August 23, an editorial began, "Greenbelt just ain't what it used to be." The lack of interest in local affairs caused great dismay. On September 25, 1952, a headline urged, "Stand Up and Be Counted." On May 20, 1954, the paper reminded everyone to register to vote in the primary election. After repeated editorials and a large turnout, the September 22, 1955, issue commented, "We sometimes wonder what sort of democracy we would have in Greenbelt if no community paper existed. Perhaps we would be like the Colmar Manor and University Park communities, where total votes cast are between 25 and 60! May this never happen here."

Throughout the difficult period of transition, when Greenbelt evolved from federal experiment to a town with a housing cooperative, the newspaper offered badly needed continuity and stability. The staff also provided perspective on the issues at hand, along with good counsel. The following editorial was especially wise, having been published on April 29, 1954, at the height of the McCarthy era:

> Some people have concluded that a country is not safe unless it is united, which is true, and that a country cannot be united unless everybody in it agrees with everybody else about everything, which is false.
>
> Controversy is one of the best features about democratic society. A society that does not permit controversy is on the way to death. When a subject is said to be controversial it means simply that it is alive and people are arguing about it. The word should not be used as a term of reproach, but rather as a compliment. Engaging in controversy is one of America's most useful as well as enlivening indoor sports.
>
> This also applies to our local organizations. The differences of opinion which take place on the city council, the GVHC board, or the GCS board, do not lead to irreparable schisms. On the contrary, they indicate that these groups of men, as individuals, are bending their energies to the solution of our mutual problems. The controversies surrounding them are better solved by taking all the viewpoints into consideration before making final decisions. This is the democratic way.

On February 17, 1955, the paper printed an editorial on the subject of repeated requests to withhold the news. Various organizations, discussing "delicate situations," had requested that reporters not put anything in the paper as "it would ruin our organization." The paper's policy was decided, however:

> Our reporters will not be unfairly encumbered with requests to withhold information which is disclosed publicly at meetings that any member or citizen may attend and hear.
>
> Also, if this stated policy encourages closed meetings (executive sessions), this newspaper will endeavor to print any information it may learn of these meetings, as part of our job to get and report legitimate news.
>
> We are asking organizations in Greenbelt to become aware of their responsibil-

ities and we will take care of our own. We are not obligated to anyone except our readers, and we prefer that no one be obligated to us—for suppressing the news.

The staff also worked to encourage the nonpartisan political process by printing on the front page the agenda of each city council meeting in advance of the meeting. Before both city and national elections, editorials urged residents to register and vote. The paper published biographies and candidate position statements before city and Greenbelt Homes board elections, in addition to discussing both sides of any issues on the ballot. New columns such as "Librarian's Notebook" about the Greenbelt Library, "The Legionnaire," covering American Legion activities, and "Our Neighbors," a long-running column by Elaine Skolnik, began in the 1950s. A series appeared in 1958 called "Twenty Years Ago." A full-page advertisement, published "in the public interest" by the paper, implored residents to patronize the local theater, which was having financial problems.

In April 1959 the paper's board of directors reluctantly made the decision to initiate a public fund-raising drive, as lack of advertising revenue continued to be a problem. Al and Elaine Skolnik organized the drive, finding volunteers in each court to canvass door-to-door. By May the volunteers had collected more than $1,400. An April 30 editorial reveals how this lifted the spirits of the staff: "We did not begin to appreciate the immensity of the reservoir of good will existing in Greenbelt for the *News Review.* Your actions on our behalf have overwhelmed us. The enthusiasm of the volunteer collectors, the generosity with which our fellow citizens of Greenbelt dipped into their pockets to support their paper—these have been a never-ending source of pleasure and wonderment to us."

When Harry Zubkoff ended his term as editor in 1960, he wrote an editorial on January 28 summarizing his views of the paper's function: "As I see it, the *News Review* is the major unifying element within the city, the force which has done more than any other single civic activity to make a city out of a housing project. It has given Greenbelters a sense of belonging, a sense of participation, and, frequently, a feeling of pride in their community." He followed with the usual appeal: "If I have anything to say at all, it is to urge more of our readers to participate in local activities, to contribute more of themselves in terms of time and talent for the benefit of their community." As to the role of the paper: "The *News Review* will have fulfilled its function when it succeeds not only in keeping its readers informed about local ac-

Advertisement for the movie *3 Brave Men*, based on
Greenbelter Abraham Chasanow's experience with the
Office of Naval Intelligence during the McCarthy era.
Courtesy Greenbelt Museum

tivities, but also in drawing its readers into these activities, each where he can
best contribute to common progress."

Major community institutions such as the newspaper and Greenbelt Consumer Services were not the only cooperatives to maintain their functions
throughout the postwar period. The Greenbelt Health Association and the
cooperative nursery school both continued their services with the active
participation of residents. When the Lanham Act no longer provided funds
for child care, Greenbelt working parents responded by forming a child care
cooperative. The spirit of cooperation thrived in the religious sphere also;
however, the form of cooperation changed. Protestants, Catholics, and Jews
all held services at this time, posting regular notices of their activities in the
paper. They did not continue with earlier plans to build an interfaith center,
as each religious group wished to construct its own building. However, the
PHA ignored their repeated requests for land. After being frustrated in individual efforts, in September 1948 church leaders united to urge the PHA to

release land for sale or rent for church sites. In January 1949 the PHA announced the granting of bids to five local groups. Members of the various denominations built quickly; St. Hugh's Catholic Church opened in November 1949, followed by Greenbelt Community Church in June 1950. The Methodists consecrated their new building in October 1955.

Church members built the first, temporary, Mowatt Memorial Methodist Church themselves, naming it after a local war hero. The *Post* noted their effort: "The spirit of early American ministers, who built churches in the wilderness, is alive in the heart of a Washington area pastor. Although the Reverend Chester J. Craig is building his church in the thriving town of Greenbelt, he is facing many of the same problems overcome by pioneer pastors. And the problems are being solved in the same fashion—by community effort." They cleared the land, purchased a surplus GI chapel, assembled and painted it.[53]

Perhaps this example spurred the members of the Jewish Community Center to construct their own building. However, progress remained slow until the "Help Build the Jewish Community Center Week End" occurred. As described by the *Washington Star:* "The idea was conceived by a Catholic, taken up by a Congregationalist, and is being carried through by followers of all three major faiths."[54] Members of Greenbelt Christian congregations took turns in construction, while Jewish groups in the greater Washington area sent aid and supplies. With very little professional help, and a number of mistakes, the building reached completion in 1955. The March 17 *News Review* commented: "Sunday's dedication is considered by residents of Greenbelt as a symbol of neighborliness and brotherhood, as well as a house of worship." This brotherhood was not just rhetoric aimed at newspaper readers. Residents remember a prevailing attitude of religious tolerance.[55]

In the postwar years, Greenbelt residents also demonstrated their ability to cooperate in the education of town children. In February 1952 the city council announced the end of kindergarten classes due to lack of funds. The state of Maryland did not mandate kindergarten, but Greenbelt had previously provided it with funds from the federal government. Concerned parents canvassed their neighborhoods, got 400 petitions signed, and presented them at the following city council meeting. The Parent-Teacher's Association sent telegrams to congressmen, urging them to intervene with the PHA for funds. None of this produced any money from the city council or the PHA, so the parents set up and ran a cooperative kindergarten, continuing until the state began kindergarten education in 1957.[56]

In the fall of 1954 the Prince George's County Board of Education opened High Point High School, located north of the National Agricultural Research Center, for students of the surrounding towns, including Greenbelt; for the first time, Greenbelt students attended school away from their town. Greenbelt High School became Greenbelt Junior High. This change was only the first in an increasing merger of Greenbelt into Prince George's County in the area of education. However, such merging did not happen easily or smoothly. The North End Elementary School remained under the jurisdiction of the Federal Works Agency, having been built with funds from the Lanham Act. The federal government sold both North End school and Greenbelt High School to the county's board of education, and, after completing necessary changes in July 1958, sold Center School to the county for $260,000. The building continued to be used as a community center also, with the city paying rent for this use.[57] Even with growth at a relatively slow rate, Greenbelt schools began to approach the limit of their capacities. Overcrowding at High Point High School forced the use of split classes, and plans were under way for an additional high school.[58]

In addition to the many cooperative efforts of a business, religious, and educational nature, local groups maintained town traditions. In 1945 the Labor Day town fair attracted 5,000 participants. Planning for the following year began almost immediately. All-Greenbelt Night, held in January, became so crowded by 1948 it had to be held at Ritchie Coliseum at the University of Maryland. Featured events that year were a double header basketball game, tumbling by elementary school children, and the crowning of the queen of Greenbelt High School. In 1949 the Fourth of July holiday featured three days of special events. In 1955 the Labor Day program became a four-day festival for the first time. The August 4 *News Review* described events to be included: "Four nights of street dancing on Centerway, which will be closed off during the festivities. In addition there will be square dancing on the tennis courts, four nights of professional entertainment, a majorette contest, fire company contests, two boxing shows, an art exhibit, a mammoth parade, a popularity contest, a fishing rodeo, athletic contests, and numerous rides, games and Bingo. No dice games or gambling will be permitted."

In June 1947 residents celebrated the tenth anniversary of their town.

Washington's newspapers noted the occasion, with the *Post*'s headline exclaiming, "Greenbelt—the Boondoggle That Made Good." A smaller headline proclaimed, "Federal Town of 8000 Nears Happy 10th Birthday." The article concluded:

> Well, Greenbelt today is a pretty nice place, with a community spirit that gets things done. The town's cooperative stores are in the black. So are the cooperative buses which, about once a month, give free rides as dividends. At meetings of the citizens' association, American Legion post, sewing circles, etc., there is plenty of outspoken griping—a sure sign of a healthy, democratic community. As Town Manager Gobbel says, inhaling a chest full of that good old Greenbelt air: "There's nothing like it."[59]

Representative Frank Buchanan entered the article into the *Congressional Record* of June 9, 1947. A *Post* reporter also noted, "Old-timers—and there were many—among the crowd of 3,500 packing lawns along the parade route seemed to have forgotten the day when this model of a planned community, operating entirely on a cooperative basis, was a fiercely opposed project decried as a 'New Deal' extravaganza."[60] The article did not mention that the *Post* had been one of the foremost critics of Greenbelt.

Soon after, on the occasion of Greendale's tenth anniversary celebration in May 1948, Rexford Tugwell gave a speech, which included these remarks:

> The *Washington Post,* like most newspapers, I suppose, has a short memory. Last year, I was amazed to receive, from numerous correspondents, copies of a Sunday issue in which Greenbelt was pictured on its decennial as rather more of a paradise than its residents know it to be. Gone were the days when it threatened the foundations of American enterprise so terribly that it had to be savaged day after day through months of campaigning. Gone were the days when its costs were so absurd as to be comical to all right-thinking men. It was now a great success, a planned institution for living which met more of people's fundamental needs than any other design in any other place or time.
> So perhaps the campaign may after all be won. As for myself, even though I lost an opportunity to be of service because I was a poor general, I notice that it is no longer considered so funny as it once was to call Greenbelt "Tugwelltown."[61]

Anticipating Tugwell's remarks, the *Washington Star* headlined its article on the tenth anniversary celebration, "Time Takes the Laugh out of That Funny Word Tugwelltown." The *Washington Daily News* joined the chorus with its banner, "Greenbelt: Where the Library Collects More Fines than the Courts." Even President Truman jumped on the bandwagon after reading the article in the *Post.* On July 31, 1947, he sent a letter addressed to the people of Greenbelt:

It's too bad the story of Greenbelt is not better known. We all recall the ridicule that was heaped upon this enterprise when it was started. Now the POST describes Greenbelt as a community that has "flourished and prospered in a manner to confound its critics." I understand you are about to celebrate your tenth anniversary—that you are going to have national speakers, a pageant, dancing in the streets and so on. All this is most fitting. On this happy occasion I want to extend to you my warm congratulations and best wishes. Very sincerely yours, Harry Truman.[62]

After the tenth anniversary celebration, city traditions continued and for the first time often occurred without the instigation of the newspaper staff. Other groups—the Boys Club in 1956 and the Fire and Rescue Squad in 1957—stepped forward to sponsor Fourth of July parades and fireworks. The *News Review* heralded the 1956 Labor Day festival as the "most spectacular ever seen." A two-hour parade climaxed four days of events, with all proceeds from the festival going to build a youth center.

The city council's campaign slogan, "We Are Cooperating," provided impetus for a spring cleanup in 1958. A number of clubs and social activities continued to thrive; the Lions Club published an updated city directory every few years to help residents keep track of each other and of their organizations. In 1957 the Golden Age Club appeared and soon had its news items published in the *News Review.* The establishment of this club, with increasing attention focused on its activities, signaled the aging of original Greenbelt residents. So many athletic and recreational groups existed that at least twenty organizations met with the city council in 1957 to form the Greenbelt Recreation Coordinating Committee. This replicated the wartime pattern of forming a committee composed of representatives of other committees or clubs to organize and coordinate their activities. By maintaining their many local groups and traditions, residents maintained their self-concept as activist members of a closely knit, cooperative community.

Other than the extensive coverage of the tenth anniversary celebration, the only attention Greenbelt received from the local media in this period was coverage of its travails with the federal government, which have been discussed. However, the story of Greenbelt began appearing around the world. The *Star Weekly* of Toronto, Canada, proclaimed "They Heard the Hungry" in August 1946:

When President Harry S. Truman first informed the United States that drastic measures were necessary to conserve grain and fats for the starving, the citizens of Greenbelt, Md., were among the first in the country to go into action. Accustomed to cooperation in every phase of their community living, they sensed immediately that the federal government's efforts would have to be supplemented by individual voluntary rationing if the president's commitments were to be met. Within a few days they had devised a food conservation program which became a model for other cities.[63]

International visitors to Greenbelt included F. J. Osborn, a founder of Welwyn, the second English garden city; architects from New Zealand, Norway, and West Germany; and students of cooperatives from Indonesia and Kenya. The individual in charge of postwar construction in Warsaw arrived to see the town's superblocks, as they were to be used in rebuilding his city. The planners of Los Alamos, New Mexico, wrote to city manager Gobbel for details on setting up a town government. The continuing interest in their town, especially from abroad, must have gratified the residents and demonstrated to them that its design and its co-ops remained unique.

In addition to celebrating their tenth anniversary and maintaining town traditions such as the Labor Day festival, Greenbelters renewed their interest in city governance. In the city council election of September 1945, 79 percent of those eligible to vote did so. Nineteen people entered a hotly contested race in 1947, running for the five city council positions. An astounding 96 percent of Greenbelt's registered voters turned out for this election.[64] The candidates generated intense interest by the genuine difference in viewpoints they represented. The winners favored cooperative ownership of the government housing, so the contest served as a referendum on the future of the town by its residents. In 1949 Greenbelt voters elected Betty Harrington for council, who became the first woman mayor in Maryland.[65] Greenbelt citizens continued to go their own way, voting heavily for Stevenson over Eisenhower in the 1956 election.[66]

Heralding a new trend, in the 1954 county election, Mayor Frank Lastner won a seat as a Prince George's County commissioner. According to the November 4 *News Review:* "He stated that his nomination and election indicated that Greenbelt was now accepted as a full participant in the affairs of Prince George's County and was no longer regarded merely as a government project isolated from the rest of the county." This trend continued, as in February 1955 George Panagoulis, chief of police in Greenbelt, became the county chief of police. After the federal government sold Greenbelt, the only option available to residents to overcome their isolation was to reach out to the county government.

In 1958 city council members proposed changes in the city charter that would have had the effect of increasing the power of the council and reducing that of the city manager. At least a portion of those on the city council wanted more power for themselves, particularly those who wished to influence council decisions in favor of more development. A group against such development, calling itself Citizens Charter Referendum Committee, promptly formed in order to request a citizen referendum on the proposed changes. Fifty volunteers went door-to-door collecting signatures on petitions requesting a vote. Voters went on to defeat the proposed changes in a citywide referendum.[67]

The actions of the youth of Greenbelt showed that they had been closely observing their elders. When the upcoming sale of Center School to the Prince George's County Board of Education curtailed their Friday evening roller skating privileges, more than fifty youngsters appeared at a city council meeting to ask for a resumption. The council worked out an agreement to pay for a new floor, and the skating resumed.[68] This success demonstrated to the teenagers that their participation in civic affairs would be rewarded.

Greenbelt residents remained active and aware not only in city government but also in matters that affected them more personally, such as rent payments. In September 1948 more than 500 residents attended a meeting initiated by the *Cooperator* to protest a rent increase. As a result, President Truman received a deluge of letters and telegrams, and the rent increase was reduced to half the amount originally scheduled. In July 1951 a rent increase of 15 percent appeared imminent. A Rent Protest Committee formed, more than 400 citizens came to a meeting, and the government delayed the increase for sixty days. In August Congress passed a law that maintained rents at the original level. In January 1953 more than 300 residents sent protests to officials of the Office of Rent Stabilization regarding a proposed rent increase and then created the Greenbelt Tenants Association to fight increased payments. The citizens of Greenbelt successfully delayed and decreased planned rental hikes throughout the period of the town's sale by holding mass meetings, forming committees, and sending letters.[69]

Throughout the postwar period Greenbelters clearly demonstrated that they could organize to achieve their goals. With aims as varied as prevention of rent increases, formation of a co-op to buy their homes, maintenance of town traditions, and the continuation of co-ops, the weekly paper, and numerous social activities, they organized to achieve their goals. In addition to the endurance of a myriad of community organizations, they turned their attention to the preservation of the physical integrity of their town. In maintaining their planned community, residents utilized their original ideals, or Greenbelt principles, as the yardstick by which to measure change. In 1946 state and federal government agencies began planning both the Baltimore-Washington Parkway and an air-freight depot adjacent to town. As early as May 10, 1946, an editorial in the *Cooperator* referred to the looming loss of the encircling greenbelt: "We like to think of the plan under which Greenbelt was originally built and wish the Federal Public Housing Authority would think of it too sometimes. A super-highway only 100 to 200 feet from Greenbelt's back door and a large air-freight depot just beyond were not part of the philosophy of the original plan." Construction of the parkway proceeded, but the depot construction was halted.

In an editorial on June 10, 1948, the paper suggested that the town create a planning commission to see that development was thoughtfully carried out. In November 1953, as buyers expressed interest in vacant land, the city council passed a resolution to keep Greenbelt a "garden city." The November 5, 1953, *Cooperator* printed the resolution in full on the front page. The key paragraph summed up the wishes of town residents:

> The objectives of the determinations of the City are: (1) to continue the town planning principles which were applied in the original development of Greenbelt; (2) to maintain the plan for a "green belt" to protect the central areas of the City which are developed in higher densities; (3) to avoid through traffic in new areas of residential development; (4) to make adequate reservations for future playground, park and school sites to serve the new residential areas; and (5) to require new developments to pay the cost of extending utilities and roads, in order to avoid City expenditures therefore in view of the necessity for achieving a reasonable City tax rate.

The article summarized the effect of the resolution: "It served notice to future developers, builders, and purchasers of undeveloped property in Greenbelt that the city will insist on zoning regulations and principles that would keep Greenbelt a 'planned garden community,' which would follow in spirit the Hale Walker plan for Greenbelt filed in the National Capital Park and Planning Commission office."

On April 28, 1955, an unhappy resident wrote a letter to the editor entitled "R.I.P":

> Here lies the planned community of Greenbelt, Maryland! The planners dreamed dreams of a shopping center where the children could enjoy a trip to the play-

Citizen's meeting in school–community center, Greenbelt,
1950s (Mayor Thomas J. Canning is speaking). Photo by
Nick Pergola

ground, while their mothers purchased the daily food supplies. In a sentimental mood, they put up a fountain-statue of the mother giving her child a drink. Visitors from all over the world raved about the sense of space, the pleasant vista, created by the little park at the edge of the shopping center. Now is our city park to become a city hall or (O thing of Beauty!) a parking lot? Shall we replace the sentimental statue, too, with something more in tune with the times—say a neon $ sign?

Development threats continued to appear, whether town officials and residents were ready or not. Greenbelt attracted the frequent attention of highway planners because the government already owned the land, thus cutting costs. The Baltimore-Washington Parkway appeared in 1948, built to the immediate southeast of homes in Greenbelt. In 1954 planners announced the route of the Inter-County Belt Freeway, drawing criticism from residents. The proposed route placed the freeway near the lake, plowing through the middle of a proposed housing area. The city council sent a resolution to Governor Theodore R. McKeldin, asking him to intervene on the city's behalf with the State Road Commission, stating that the road would interfere with the orderly development of Greenbelt as a "planned community." Other factors involved were "the destruction of recreation areas and historical Indian Spring, interference with the city's sewage disposal system, and the possibilities of new traffic hazards which would jeopardize the municipality's fine traffic record."[70] The planners accepted an alternate route, not because of the pleas from Greenbelt but because a cemetery stood in the path of the projected roadway.

These two highways formed immense barriers—freeway canyons—between original Greenbelt and the residential areas built later. If the homes had existed first and relationships were already formed before the barriers were put in place, it might have been possible to keep connections with these outer sections. As it occurred, with the barriers first and newer neighborhoods afterward, it proved an almost insurmountable challenge to form a cohesive Greenbelt.

Initially, new neighborhoods formed within the freeway barrier, in space adjacent to the original housing. After the sale of the town, many Greenbelters wished to resume their efforts, which had been interrupted by the war, to form cooperatives to build larger homes. First, the GVHC itself hired a private builder to develop the available land. However, the Federal Housing Administration stymied these efforts by refusing to underwrite mortgages on the type of homes planned. The FHA contended that the homes would not be saleable because Greenbelt's taxes were too high (in fact, the highest in the area; Greenbelt residents taxed themselves in order to continue the services no longer paid for by the federal government).[71] After the government rebuff, a group of residents organized to buy twenty-five acres from the GVHC in a desirable location along the lake. Lakeside Homeowners, Incorporated, formed as a cooperative to buy lots for members and then sold the homes to individuals. The group worked quickly, with the first home finished by December 1954. Greenbelt, long a community of equality, began to be stratified, as some of the more financially well-off moved to Lakeside.

In 1953 the GVHC created Greenbelt Community Builders, Incorporated, a nonprofit cooperative housing venture. The cooperative purchased land and built homes, which it then sold to individuals. They produced Woodland Homes, a 50-unit cooperative of two-, three-, four-, and five-bedroom houses; and Ridgewood Homes, 104 three-bedroom houses. Together, these form the area now known as Woodland Hills, located on an extension of Northway, adjacent to the old defense homes. A representative of the Foundation for Cooperative Housing, James J. Cassels, came to assist Greenbelt Community Builders and stayed to purchase a home for himself. On a business trip to Chicago, he discussed the new homes with Rexford Tugwell, who was making plans for his retirement from the University of Chicago. Tugwell decided to purchase one also, as he wished to do research at the Library of Congress and evidently felt that living in Greenbelt was an attractive idea.[72]

As the interest of Cassels and Tugwell show, people other than those living in GVHC housing moved into the new homes of Lakeside and Woodland Hills. However, previous Greenbelters dominated both areas. They brought with them their ideas about the community, its traditions, and history. Thus the gospel according to Greenbelt spread out from the original housing to new areas of town.

What can be seen in retrospect as a serious error on the part of the GVHC management occurred at this time. The GVHC continued to have difficulties not only with internal squabbles but also with managing the intricacies of high finance. It was, after all, an organization of volunteers, composed largely of government employees with no special knowledge of land management. In 1955 they found themselves lacking sufficient funds to pay taxes on the undeveloped land, and rather than seek financing to retain the land, the board of directors decided to sell it. They saw further development as a means of increasing the tax base and keeping city taxes at a reasonable level. In May 1955 the GVHC sold all the land but that committed to Greenbelt Community Builders to the firm of Warner-Kanter. Interestingly, this firm served as the developer of Forest Park, created from the land of Greenhills, Ohio.[73]

By this sale to a private developer, the GVHC allowed control of the future of the town slip from its own members into the hands of outsiders. Because Greenbelt residents wished their town to develop according to the Greenbelt principles, they needed to maintain control over the vacant land to ensure that their wishes would be followed. The Prince George's County government determined zoning laws, not the city of Greenbelt, thus making it even more difficult for Greenbelters to focus development in the manner they desired. However, town residents had not yet realized these hazards.

In the decade of the 1950s, Greenbelt and its residents survived harsh trials, forging themselves into a more defined community as a result. In the process of working together to save their town, Greenbelters reinforced their communal values, traditions, and organizations. For the first time, people in Greenbelt lived there because they had chosen it for their long-term future. In this period, not only was the physical core of Greenbelt maintained, but it became a cooperative. Housing cooperatives formed and built new housing developments, largely populated by Greenbelt residents, who thus spread the ideals of their planned community beyond the original borders.

The cooperative businesses continued to thrive, even expanding beyond the town. Some viewed this success negatively, as, in the view of many, Greenbelt Consumer Services had become just one more big business. However, the success and growth of the GCS could be seen as an affirmation of cooperative principles. Other forms of cooperation continued to manifest themselves, as the town's many social, religious, city improvement, athletic, and educational groups flourished. The city newspaper continued, and town traditions, especially the Labor Day fair, proceeded as usual.

With the formation of the housing co-op, a core group of activists emerged who worked together to ensure the long-term well-being of their community. Other activists continued to publish the weekly paper, while those with a special interest in politics ran for the city council. These town leaders did not hesitate to ask for the support of the larger community whenever necessary. The continuing crises served to tie people together as they relied on each other for support. Thus even though the postwar period posed many challenges, the community of Greenbelt emerged with its original goals and ideals not only intact but even spreading.

The quiet time of the late 1950s provided Greenbelters with a much-needed respite. After the disturbances caused by the sale of the town and the Chasanow case, Greenbelters needed time to rest and to pull themselves together. It proved particularly fortunate that they had this period of tranquility, as they were soon plunged into renewed turmoil when their one mistake caught up with them.

III

# Threats to Greenbelt's Plan and Cooperation

# 5

# Developers versus Greenbelt

DURING THE 1950S Greenbelt residents adjusted to life in their suburb without the presence of the federal government. They continued to perceive Greenbelt as unique and resolved to maintain the original goals of the town on their own. During the 1960s this resolve faced severe tests when developers began construction on Greenbelt's vacant land, as these entrepreneurs felt no inclination to adhere to the Greenbelt principles. When residents realized they faced an uncontrolled building boom in their midst, they fought back with every resource at their command, continuously invoking the legacy of the past. How did Greenbelters remain cohesive and ready to protect their community, even as their difficulties piled up, one after another, year after weary year? Their concept of Greenbelt as a planned cooperative community remained stable, guiding their behavior even as the physical reality of the town changed around them.

The first privately owned and financed construction in town was built by J. Evans Buchanan and Company on land adjacent to Woodland Hills, which was purchased from Warner-Kanter. It consisted of 102 homes and became known as Lakewood. As in Woodland Hills, this area contained many young and growing families from Greenbelt Homes, Incorporated, who wished to stay in Greenbelt but wanted single-family homes.[1] Warner-Kanter soon sold its remaining 560 acres, with ownership changing many times. The Dan Ostow–Charles Bresler Corporation built the Greenbelt Plaza apartments on one parcel in 1959. These two-story apartments, located centrally, right behind the shopping center, fit in with the city's building style and plan. Up to this point, new building occurred in areas adjacent to that already developed.

Even though the federal government no longer owned Greenbelt, it remained involved in several ways, contributing to its growth. In 1958 the National Park Service announced a plan to develop Greenbelt Park into a 1,100-acre recreation area, with public campsites and an eighteen-hole golf course.[2] When the National Aeronautics and Space Administration came into being, also in 1958, the federal government located it on 500 acres of land belonging to the Beltsville Agricultural Research Station. Senator J. Glenn Beall announced, "Greenbelt is an ideal location for the new government agency because of its accessibility to the nation's capital and its proximity to numerous important highways."[3] Original plans were for a workforce of 450, a number that rapidly grew. Many employees settled in Greenbelt, stimulating a significant demand for housing and city services. In 1959 a bond issue passed to finance a youth center and a building to house the fire department and the rescue squad.

In November 1956 the Maryland–National Capital Park and Planning Commission (MNCPPC) drew up an overall plan for growth and development, "A General Plan for the Maryland-Washington Regional District." The plan describes Greenbelt as "an example of a residential community that embodies many of the desirable features" of good residential communities. These features were shopping facilities within a reasonable distance, walking distance if possible; an internally located elementary school, with one or more playgrounds; a street pattern that provides easy circulation within the area but that discourages through traffic; separation of people from vehicular traffic, so that pedestrians can walk to school, playground, or shopping center without crossing major streets; building lots arranged in a compact manner, with suitable open spaces; and preservation of trees and wooded areas.

Greenbelt may fall short of present-day standards of housing design, and the row house may not be the dwelling type now most in demand, but . . . in many respects it is still, after 20 years, the best example of a suburban community designed for the automobile age. In such a neighborhood, it is possible to live in comfort and safety. Children can walk to school and playground without crossing busy streets; it is not necessary to drive several miles to buy a loaf of bread; the noise of traffic and commercial activity are removed from the residential area. Such a carefully designed community also offers savings over a poorly designed community—less land is needed for streets, and public utility lines can be shorter.[4]

However, ironies abounded. Although the MNCPPC report extravagantly praised Greenbelt and its design, it contained the seeds of Greenbelt's destruction. One proposal called for "redevelopment" of the "temporary" housing area, replacing this housing with single-family, semidetached, and apartment housing. The population forecast was for 12,000–16,000 people—considerably more than the current 7,000—with most new housing to be in high-rise apartment buildings. The plan called for a twenty-acre shopping center at Greenbelt and Edmonston Roads, with a new high-rise apartment building across the street. It also called for two new commercial buildings in the current shopping center, with a new street encircling this area.[5]

Because Greenbelters maintained their original vision of their town, based on Hale Walker's plans from the early 1950s but with increased emphasis on single-family housing, they disagreed with most aspects of the MNCPPC plan. They felt the new shopping center unnecessary, as additional facilities could, and should, be built on vacant land near the current shopping area.[6] More than forty Greenbelt residents who strongly objected to the MNCPPC plans attended a hearing at Center School on February 28, 1957, expressing their views that the proposed high-rise apartments and huge shopping area in the Center would irreversibly change the character of Greenbelt, making a cooperative community virtually impossible.

The conflict between Greenbelters and developers simmered in the background until the building of the Capital Beltway in the early 1960s, which spurred development by improving accessibility to Greenbelt. The building boom of the 1960s, with its challenge to their way of life, provided residents with the most difficult tests they had yet faced. Developers threatened the very existence of the town because they enjoyed the cooperation of county and state government bureaucrats, planning officials, and the judicial system at all levels, thus pitting Greenbelters against a formidable array of opponents. Only the overwhelming desire to maintain the physical integrity of their town kept residents working together throughout the decade.

Greenbelt residents did not simply react against development but rather worked actively for a particular kind of development they felt would enhance their community. The universally agreed-on plans remained those espousing the Greenbelt principles. Residents wished to maintain the original tenets of Greenbelt's design, with emphasis on the neighborhood unit and a walking community, surrounded by a belt of green. Because the government-created housing consisted of town houses and apartment buildings, residents wished to see new development in the form of single-family homes, as had been done by cooperative groups in Woodland Hills and Lakeside. So while residents fought against the plans of developers, they also fought for their own vision of what Greenbelt should be. Greenbelters lived in a planned community, and they wished to keep it that way. An advisory planning board of citizens began work on February 16, 1961, to discuss the form future development should take.[7]

The year 1962 saw the entrance on the scene of Charles Bresler, a developer who would have a significant impact on Greenbelt and its residents. On April 16, 1962, Bresler signed several covenants with the city setting forth conditions for his 50-acre development of town houses. Bresler had requested R-18 zoning, with twenty-one units on an acre—which the MNCPPC did not recommend. In a deal made with the city, Bresler donated 3.3 acres of his land for parkland and agreed to limit construction to seven units an acre. The city council accepted the covenants and agreed to recommend approval of Bresler's rezoning application to the county commissioners, "with an accompanying statement that the proposed development conforms to the planned community concept on which Greenbelt is based."[8]

On June 29, the county commissioners approved Bresler's application for R-18 zoning, presumably because of the recommendation by the Greenbelt City Council. Bresler planned to build 350 town houses, with four to seven buildings grouped around a central square.[9] Following this approval, Bresler met informally with the city council to discuss land planning. He suggested that a professional planner be employed jointly by his company and the city to work out a plan that would meet with everyone's approval. At the time, he and his partners owned 350 acres of vacant land in Greenbelt. Bresler said that he wanted to develop the land in accordance with city wishes, otherwise "there will be constant bickering and waste of time in zoning controversies."[10] The council planned to meet further with Bresler representatives and the city's Advisory Planning Board.

After four months of discussion and a concrete proposal, an editorial in the November 15 issue of the *Greenbelt News Review* urged caution: "Greenbelt . . . stands on the threshold of an enormous building boom. Various private developers have expressed an interest in this area. Because of this situation, the *News Review* feels increasing concern over what appears to be a potentially dangerous entanglement of municipal interests with those of one private builder, Charles Bresler, owner of most of the remaining undeveloped land in the city." The citizens' Advisory Planning Board objected to the city joining with Bresler to pay $8,000 for a new master plan for the city for two reasons: they felt the landowner should plan his own objectives, and much of the planning for the city had already been done by the MNCPPC. The newspaper brought up another point:

> Of even greater concern is the conflict-of-interest problem inherent in the Bresler proposal. The planning firm to be hired is one that has been in Bresler's employ for some time. And although we feel certain that Bresler's own motives are generous and above-board in offering this joint proposal, we cannot believe that anyone can successfully serve two masters. We feel it is impossible to ask that a planner neatly combine the best interests of both the community and its major private developer.

The city rejected Bresler's planning offer.

The year 1962 ended with the opening of the first portion of "the Washington area's first climate controlled, completely covered shopping center," heralding the beginning of large-scale development in Greenbelt.[11] Built by developer Sidney Brown, Beltway Plaza appeared on seventy acres along Greenbelt Road west of the Beltway.

In 1963 Community Builders, Incorporated, began to construct Springhill Lake, described as the largest garden apartment complex on the East Coast by Jim Giese, the city manager. The developers studied Greenbelt's design and largely replicated it in their 3,000-unit plan. Apartments with ample green space surrounded a central shopping area. A swimming pool and a system of internal walkways were included. This development caused no outcry or protest by residents as it followed the Greenbelt principles fairly closely.[12]

The issue of development dominated Greenbelt city government in 1963, with many residents seeing the city council as too prodevelopment. On September 12, 1963, 150 residents attended "Meet the Candidates Night." Mayor Francis White, Ben Goldfaden, and Bill Phillips formed the prodevelopment majority. Dave Champion presented himself as a "minority" against development, while Dick Pilski defended the council but declared himself "independent" of the others. Seven candidates challenged the incumbents, focusing particularly on the prodevelopment group's adversarial relationship with the Advisory Planning Board. Lewis Bernstein, one of the challengers and the chairman of the APB, stated that the treatment the board received from the city council had made him angry enough to run for election. The candidates' forum took place when the city council had under discussion the joint hiring of a city planner with developer Charles Bresler.[13]

Mrs. James A. Halsted, daughter of President and Mrs.
Franklin Roosevelt, at ceremony honoring her mother,
Greenbelt, November 2, 1968. Photo by Borland
Photography, courtesy *Greenbelt News Review*

Eighty-one percent of registered voters went to the polls for the city election on September 19, 1963. To win a council seat a contestant needed a
majority of 698 votes. Only two candidates succeeded: Edgar Smith, a former
councilman who had criticized the council as a "complete failure," and Dave
Champion, the one council member against development. The top six finish-
ers took part in a subsequent runoff election. The three current prodevelopment council members received the fewest votes, thus being replaced by three
challengers. The September 26 issue of the *News Review* commented, "The
newly elected councilmen ran on a program of preserving Greenbelt as a
residential community, providing suitable controls over developers, and encouraging citizen participation through open council meetings and citizen
advisory boards." This election illustrates the process through which Greenbelters ensured that their elected representatives reflected the majority views
of residents.

The January 2, 1964, *News Review* provided an excellent summary of the workings of the community:

> It would be a fair statement to say that 1963 was not a quiet year in our community. The tides of controversy swept through the town in seeming constant succession. At the same time, the controversies that raged were conducted in the traditional Greenbelt grand style. Issues were debated heatedly, with letters to the *News Review* flying thick and furiously, and then the arguments were resolved by the ballot box or by other reasonable and democratic means. The battle then abated as the participants relaxed and waited for the next issue, which is bound to come along any minute.

The editorial also demonstrated the way in which Greenbelters perceived themselves: "Again we saw working the old Greenbelt adage—put two Greenbelters together and you can get an argument started, put three together and you have a new organization."

> Did all this ferment in 1963 prove anything? We think it did. It demonstrated that Greenbelt's institutions are strong and can endure the stretching and pulling of factional feuds without permanent harm. It reinforced the foundation of our community as an almost unique place where the old-fashioned type of "town hall meeting" atmosphere is always available for people to get things off their chest and speak their mind on almost any subject. On this basis we can look forward to 1964 with confidence that we will see another exciting stimulating year in our history.

In 1964, with the opening of the Beltway, the pace of development increased. It now took only five minutes to reach downtown Greenbelt from the Kenilworth Road exit. Charles Bresler began construction on several sites. Boxwood Village, a section of individual family homes, soon appeared adjacent to the Lakewood area, attracting many families from old Greenbelt who wished to move out of their town houses but remain in Greenbelt. This completed the area of single-family homes north of the lake made up of Woodland Hills, Lakewood, and Boxwood Village. Bresler also built Lakeside North, an apartment complex adjacent to Boxwood Village. He began work on Charlestowne North and Charlestowne Village, apartments adjacent to his University Square. Charlestowne North was the first high-rise apartment building to alter the skyline of Greenbelt.[14]

In January 1964 the MNCPPC released staff studies that projected a population of 50,000 people for Greenbelt. (See table A.1 for actual figures.) The plan provided for 13,000 dwelling units, of which 80 percent were apartments.[15] Reaction to this appeared in a March 5 *News Review* editorial entitled "Where Is Greenbelt Going?":

> As more and more of the plans for the vacant land in Greenbelt come off the drawing boards, it appears obvious to us that our concept of Greenbelt as a planned community differs widely from that of the developers. Our concept, and one that we believe is shared by the majority of Greenbelt's citizens and officials, visualizes the future Greenbelt as a balanced community. There would be some apartment dwellings, some commercial zoning, some low-cost housing, some medium-priced individual and duplex housing, and some high-quality free-standing homes.

After giving the developers' side of the story, which pointed out the benefit of decreased taxes, the paper asks,

> Has anyone asked whether such a metropolis is what Greenbelt residents want? Has anyone asked whether Greenbelt residents are so desirous of tax relief that they are ready to turn the rest of Greenbelt into a community like Langley Park, consisting primarily of commercial strips and rental apartments? . . . We think the developers would be well-advised to consult once again the Community Goals adopted by the city council on November 25, 1963. This document, prepared by the Advisory Planning Board as a guide for the future development of the city, calls for the preservation of Greenbelt as a residential community and for development to the maximum of individually owned homes or apartments.

The paper concluded with a forecast that proved accurate: "It can safely be predicted that failure to heed these objectives will find the developers faced with constant, endless opposition and controversy along every step of the way."

In June 1964 the city hired both an attorney and a planner for the purpose of fighting the newly released MNCPPC master plan, which, in addition to high-density development, showed four-lane major highways bursting through the center of town. At this time more than fifty land-rezoning petitions from speculators and developers had been filed on city land. On July 21 the ten-member MNCPPC held a public hearing in Greenbelt on what it stressed was only a "preliminary" plan. More than 600 Greenbelters jammed Center School auditorium to hear witnesses testify. The Washington planner George T. Marcou, hired by the city, pointed out that the proposed increased housing density "would create in time an entirely different community and destroy forever the open, uncrowded, safe and pleasant character of Greenbelt that has become its hallmark."[16]

Greenbelt mayor Edgar Smith charged that the master plan would create "a gray belt instead of a green belt." John Gibson, a member of the city's Advisory Planning Board, emphasized that the only way "to have a green belt if the master plan is adopted would be to paint the apartment roofs green."

The attorney hired by the city, Joseph DePaul, stressed, "This is not simply a matter of a plan. . . . This is a matter of existence." The Greenbelt Junior Chamber of Commerce presented petitions with 1,476 signatures, gathered in just six days, opposing the MNCPPC master plan and supporting Greenbelt's own plan.

In February the MNCPPC revealed its unchanged final plan at an unannounced meeting and quickly forwarded it to the county commissioners, thus allowing no public comment. The Greenbelt City Council went on the attack in March, adopting its own master plan, which emphasized the preservation of balance between population density and open space.[17] In September, at the dedication of Greenbelt's new municipal building, speakers reminded their audience of the town's unique planning history, which they felt must be retained.

Some Greenbelters considered the adoption of their own master plan to be simply the first step. The president of Greenbelt Homes, Charles Schwan, called together a group, including representatives of the Greenbelt City Council, the Greenbelt Citizens Association, the Businessmen's Association, the Lions Club, Twin Pines Savings and Loan, and "leading private citizens." Schwan declared that "Greenbelt stands at the crossroads at as crucial a time as any in the history of the city. . . . Adverse court zoning decisions and the proposed Area 13 plan for Greenbelt threaten to change the whole character of the community." The committee agreed to form an organization, called Citizens for a Planned Greenbelt, dedicated to promoting Greenbelt's master plan.[18]

The April 8 edition of the *News Review* announced a door-to-door membership drive for the new organization. The steering committee announced the goal of the organization: "To preserve the fundamental character of Greenbelt as a low-density planned residential community as spelled out in the city's official goals and to insure that the future growth and development of the city should be in accord with the Greenbelt Master Plan." The group issued membership cards bearing this statement and the purpose of the group, which was "constant and firm pressure directed at elected and appointed officials of the county, state, and federal government. The CFPG will provide citizen representation at all hearings affecting the development of Greenbelt and take legal action when required." It also worked to keep residents fully informed of events as they occurred and to tell them of actions that needed to be taken.[19]

In 1965 Bresler completed his development of Boxwood Village, which contained 240 homes. A group of residents soon took court action against him, claiming that he had not remedied certain deficiencies in the development. Others expressed unhappiness over the fact that, although Bresler had promised that single-family houses would be built on an adjacent undeveloped ten acres, he had recently petitioned for its commercial use.[20]

A long-running controversy involving Bresler came to a head late in 1965. In February 1962 the city manager had conveyed to the Prince George's County Board of Education the desire of Greenbelt residents for additional schools. In March 1962 the school board began consideration of sites in Greenbelt. In October 1963 the board informed the city of a proposal received from Bresler in which he would sell land for a senior high school in parcels 1 and 2 at a certain price if there would be no opposition on the part of the city to a request for zoning land that he owned at a higher density than appeared on the Greenbelt master plan. (Parcels 1 and 2 are located between Greenbelt Homes houses and the Baltimore-Washington Parkway.) Because of the negative reception this plan received from Greenbelt residents, the city council rejected it. However, Bresler made a new proposal to the board of education to sell fifty-five acres of land in parcels 1 and 2 for a three-school complex. Greenbelt residents preferred an earlier school board proposal for a high school to be built between the Beltway and the lake, near the American Legion post. After further consideration, the school board made its preference for Bresler's plan known.

In October 1965 the city council conducted hearings on the board of education's plan; more than 150 residents attended. Jim Giese, the city manager, said that rejection of the proposal would delay the acquisition of school sites and thus delay badly needed schools but that compliance would mean agreeing to a zoning density the community did not want. Many residents spoke against the agreement, with Al Herling summarizing for the community by suggesting a series of steps that should be followed: fight Bresler's blackmail; make clear to the board of education that there would be no deals; make clear to the district council (the zoning authority) the unanimous opposition to the requested R-30 zoning; and seek the swiftest possible court settlement. "Anything less would be other than what you believe."[21]

In November the city council informed the board of education of the city's continued preference for the "beltway-lake" site. On November 30, without notifying the city council, the board decided to purchase the sites in parcels 1

Capital Office Park, Greenbelt, June 1983. Artist's rendering
courtesy *Greenbelt News Review*

and 2. In December, Citizens for a Planned Greenbelt requested that the board use these parcels for an elementary school and a junior high school and the lake site for a senior high school. The board, however, decided to use parcel 2 for all three schools, again without informing the city council.[22]

After further agitation by Greenbelt residents, the superintendent of the board of education sent a letter to the mayor of Greenbelt, responding to criticism that the board had not discussed plans with town officials. Since all parties anxiously awaited the new schools, the superintendent suggested that the city donate five acres of land from its landfill, adjacent to parcels 1 and 2; that Greenbelt Homes sell ten acres of its land for an elementary school site; that Bresler and associates sell ten acres of parcel 2 land, which would be added to the city's five acres for a junior high school site; that Bresler sell twenty-five acres of parcel 15 land (the lake site) for a senior high school; that the city purchase from Bresler and give to the board of education five acres of parkland to be a part of the senior high school site.[23] This appeared to be a compromise position, but it got tangled with other proposals before any action could be taken.

In 1966 Greenbelt residents experienced what many felt was the most discouraging year in their history. The maze of land deals between the city, Greenbelt Homes, Charles Bresler, and the school board was only one situation among many that Citizens for a Planned Greenbelt actively monitored.

In April, as development of Charlestowne Village proceeded, Bresler stunned residents by unilaterally voiding the 1962 covenants on this land that he had made with the city. He had agreed to keep density to seven units per acre, include no high-rises, and deed 3.3 acres near the lake for parkland. On June 2, 1964, Bresler filed a "Release of Covenants" with the court, without informing the city. On May 17, 1965, in an appearance before the city council, he denied reports that he planned a 400-unit development in violation of the covenants. In April 1966 an eight-story high-rise, with the number of units allowed by zoning but three times the number agreed to by the covenants, began to appear. The MNCPPC had agreed to the changes without the knowledge of the city.[24]

Al Skolnik, president of the board of the *Greenbelt News Review,* wrote a detailed story of the history of the covenants for the paper's front page. Another article covered possible city council action and the decision to sue Bresler for violating the covenants.[25] Citizens for a Planned Greenbelt leapt into action, writing letters to the editor of the Washington papers, going door to door in town recruiting members, and asking for donations for a legal fund. At their frequent meetings they attempted to monitor the rapidly changing situation.[26]

In June 1966, Bresler, a Republican delegate in the Maryland House of Delegates representing Montgomery County, announced his candidacy for the position of state comptroller. (He lost in the September primary to Louis Goldstein.) On July 18 he filed suit in Prince George's County Circuit Court asking $2 million in damages against the *Greenbelt News Review* and its president, Alfred M. Skolnik. The suit quoted three news stories, two headlines, three letters to the editor, and two editorials over a nine-month period as being printed "without reasonable justification or excuse." The quoted material dealt with Bresler's negotiations and transactions with the city council and the school board regarding his holdings in Greenbelt. Bresler accused the paper of publishing material that exposed him to "public scorn, hatred, contempt, disgrace and ridicule" and "degraded him in the esteem or opinion of the residents of the City of Greenbelt and the State of Maryland and elsewhere." The suit also claimed that the *News Review* published material specifically for the purpose of imputing to Bresler "the commission of the crime of blackmail" and "corrupt and dishonest motives in his dealings with City, County and State officials and others." He claimed that these articles caused injury to his business reputation. The Greenbelt realtor and attorney

Abraham Chasanow represented Bresler in all his suits involving the city and newspaper.[27]

On June 23 the paper published an editorial titled, "While Nero Fiddles . . . Greenbelt Burns." It pointedly questioned the city leadership for not responding effectively to the destruction of Greenbelt by developers: "What's wrong with Greenbelt's leaders? Is it indifference? Incompetence? Something else?" The city council discussed Fourth of July refreshment stands while twenty-five acres of Greenbelt went up in smoke, as bulldozers cleared land for apartments. "Still they fiddle on, but their constituents are burning." On July 7 a scathing, but seemingly accurate, commentary on the development of land in Greenbelt appeared in the *News Review*, called, "How to Get Rich Quick." After buying a parcel of land with a mortgage so the developer does not have to use his own money, he petitions county authorities to rezone the land for high-density construction. He can expect local opposition, so

that's the time when you have to be nice to everybody. Tell the city that you do not intend to make full use of the new zoning and that you will sign covenants limiting development to a lower density. Do not hesitate to sign these covenants wording them as strongly as the townspeople wish. For good measure you may even throw in a promise to deed parkland to the city. Just make sure the mortgage is all sewed up before you sign the pledge.

Now that you've got the city's support for your petition and the county authorities have reluctantly approved it, do not make the error of paying off your mortgage, even when you can do so. For if you did pay it off, you'd become sole owner and as such you'd have remained stuck with the covenants.

Instead, go easy on those payments. As a matter of fact, why not stop them altogether? The bank will foreclose your mortgage and auction the tract to the highest bidder. In doing so it can claim that it doesn't know anything of the covenants which, after all, did not exist when it granted the mortgage. And since the property now carries a higher zoning category it is bound to fetch a nice price at the sale. Such as $750,000.

And that's when everybody gets paid off. The bank takes back the amount of mortgage that is owed to it, plus its fee for the foreclosure—five percent or thereabouts. The rest is handed back to you—including about half a million dollars that's pure profit. This is your cue to easy wealth: Be audacious and presumptuous, but as far from here as can be.

In the view of Greenbelters, the developers and bankers as well as officials of agencies from the MNCPPC to the school board were out to get them, with no relief in sight. Those who studied Prince George's County during this period attest to the accuracy of this perception. Leonard Downie, in his book *Mortgage on America,* focused on Prince George's County as a place "that fell victim to perhaps unmatched exploitation of land, people, and the environ-ment during a decade of overwhelming population growth in the 1960's."[28]

Jesse S. Baggett, for sixteen years a key member of the Prince George's County Board of Commissioners, ran a political machine that provided favorable zoning decisions to developers and builders in return for sizable bribes. According to Downie, a "pattern of corruption, conflicts of interest, and influence peddling [existed] in the wholesale rezoning and development of Prince George's County." A small group of people served in multiple roles, as commissioners, real estate investors, zoning lawyers, and officers of saving and loan associations and local banks. In this speculative boom, the county population doubled, rising from 350,000 in 1960 to 700,000 in 1970. Downie summarized the result in the county by 1973:

> Fortunes were made and the face of the county changed in a generation. . . . The potential for decent living was ruined. Today, the county's streets are congested with traffic and dangerous to travel, regular public transportation does not serve most of the county, nothing can be reached on foot, and sidewalks themselves are nonexistent in places. Schools are still overcrowded in some neighborhoods, there are far too few parks and recreational facilities, raw sewage from overloaded sewers and septic tanks flows into streams and rivers, and poorly graded denuded soil has eroded so badly that some parts of the county look like abandoned strip-mining sites. Housing just a decade or two old is deteriorating into slums.[29]

This is exactly the nightmare that residents of Greenbelt could see coming to their town and that they vowed, against the odds, to resist.

Greenbelters now added the suit against the newspaper to their mounting battles. A Freedom of the Press Committee formed shortly after Bresler announced his suit against the *News Review,* to help defray the legal costs the paper would incur. In August the committee announced a door-to-door fund drive. Fifteen area captains headed the drive in the various blocks and sub-divisions, each working with a team of collectors, with all contributions deposited in a special fund in Twin Pines Savings and Loan.[30]

In September the Prince George's County Circuit Court judge Roscoe H. Parker ruled that the covenants that Bresler signed regarding the fifty-eight-acre Charlestown tract could not be enforced.[31] Bresler now planned to build at a density of twenty-one units an acre, although his agreement with the city called for seven units an acre. The city council hired a Baltimore law firm to review the covenants issue to determine if the court ruling could be appealed. The city did appeal, but in December 1967 the Maryland Court of Appeals, while ruling that the covenants with Bresler were enforceable, said the city had forfeited the right to enforce them by failing to file a complaint within the thirty days allowed in the contract.[32]

In late September 1966 the *News Review,* through its attorneys, denied Bresler's allegations stemming from his lawsuit. Their reply declared that Bresler "is both a public official and an active participant in matters of pressing public concern to which the articles in issue were directed, so that such publications are constitutionally protected and nonactionable under the provisions of the First Amendment to the Constitution of the United States and Article 43 of the Constitution of the State of Maryland." The reply also provided two additional bases upon which the publications complained of should be viewed as privileged. First, they were "fair reports of official public and judicial proceedings and actions"; second, "they constituted fair comment . . . upon matters of public interest."[33]

In October, at a hearing before the county commissioners, hard-pressed Greenbelters began to see results, as a reversal of the thrust toward high-density development finally began to occur. County commissioners began to question the massive development that their actions had allowed. Crescent Leasing Corporation asked for commercial zoning of parcel 8, located at the intersection of Kenilworth Avenue and Crescent Road, to build a small neighborhood shopping center. Charles Bresler and associates wanted adjacent parcel 7 also zoned for commercial development, both parcels being zoned at that time for single-family homes. Greenbelt's attorney, planner, and mayor, along with Greenbelt Homes, Citizens for a Planned Greenbelt, and residents of adjacent Boxwood Village, voiced their desire to keep the current zoning. The commissioners unanimously denied the zoning changes. Within two weeks of this decision, Bresler leveled the trees on parcel 7 anyway, and the owners of both parcels made several more attempts to have the land rezoned, to no avail. Eventually the city purchased both parcels to provide a scenic entrance to the community, using open-space grant money from the Department of Housing and Urban Development.[34]

By October, the city council, apparently in response to local sentiment, became more aggressive in its actions against developers. It voted to appeal a zoning approval in the Golden Triangle, the area bounded by Greenbelt Road, the Baltimore-Washington Parkway, and the Beltway. In December the council, as soon as it learned that building permits had been issued, quickly convened a special meeting and authorized its attorney to sue for "injunctive

Maryland Trade Center, Greenbelt, 1980s. Photo courtesy
*Greenbelt News Review*

relief" against construction of twenty-one units an acre at Charlestowne Village.

The year 1967 opened with further developments in the *News Review* lawsuit. Through its attorneys, the paper filed a notice of deposition, requesting that Bresler produce documents regarding his negotiations with the county board of education and the city on the site-selection process for schools as well as documents relating to financial arrangements involving the Charlestowne Village tract. Bresler's attorney, Abraham Chasanow, replied that such inquiries were "immaterial and irrelevant to the issues" and were "intended only for the purpose of harassing and oppressing" Bresler. The paper's attorneys alleged that the financial documents related directly to Bresler's claim that he suffered financial injury. A hearing on the matter was scheduled for January 26.[35]

In January Maryland's governor-elect Spiro Agnew named Bresler to the newly created post of national relations officer, to serve as the governor's liaison with state and federal agencies regarding matters in Montgomery and Prince George's Counties. At the next session of the city council several people spoke up in protest, asking for a special meeting to discuss the issue. More than 500 people met on January 30, at which time the city council passed a resolution asking Governor Agnew to "reconsider and rescind" the Bresler appointment and urged those in attendance to write to the governor. In February a seven-man delegation of concerned members of the Maryland General Assembly met with Agnew. A week later Agnew announced that Bresler would not be serving as liaison.

At a court hearing on February 23, after Chasanow agreed that Bresler suffered no pecuniary loss as the result of the *News Review* articles, the judge excused Bresler from producing certain financial records. However, he ordered that a number of other records be given to Roger A. Clark, of the Washington firm of Royall, Koegel, Rogers, and Wells, representing the paper.

The Greenbelt Freedom of the Press Committee, chaired by Charles Schwan and Al Herling, planned a number of fund-raising events to pay the paper's legal fees. The June 1 issue of the paper carried a prominent front-page advertisement paid for by the committee, proclaiming in bold print: "Don't let anyone stop the presses. Remember it's YOUR paper. Come to the kick-off rally of the Greenbelt Freedom of the Press Committee, Center Mall." State senators and delegates, as well as local officials, spoke. The former editor Harry Zubkoff pointed out that dangers to freedom of the press can take many forms, such as "expensive law suits which tax to the maximum the limited financial resources of a small newspaper. Such law suits are also used to intimidate and frighten the press and to discourage volunteers from working on the paper." The rally met its goal, obtaining volunteers for a second door-to-door fund drive, as seventy-five persons came forward to help.[36] They made detailed plans to canvass one hundred Greenbelt Homes courts, Lakeside, Lakewood, Woodland Hills, and Boxwood Homes.

During this period of fund-raising and pretrial discovery in the newspaper suit, efforts by residents to "keep Greenbelt green" continued. Resolution 76, introduced in the Maryland State Senate on March 27, 1967, by Senator Edward Conroy, gave a brief history of the city and concluded: "Resolved by the Senate of Maryland, that the purposes for which the City of Greenbelt were established are hereby reaffirmed, restated and reemphasized and that any change in this purpose is hereby deemed contrary to the original purposes; and be it further resolved that copies of this Resolution are sent to the Mayor and City Council of Greenbelt."[37] Unfortunately, good words did not always translate into favorable action. In October the Maryland Court of Appeals affirmed the commercial zoning plan for the Golden Triangle.[38]

In September 1967 the legal maneuvering regarding the newspaper suit continued. Attorneys for the *News Review* made a motion to dismiss the suit because Bresler had not released financial records regarding the Charlestowne tract. Heard in Prince George's County Circuit Court on September 19, the judge responded that one of the articles, an editorial of July 7, 1966, which referred to the Charlestowne Tract, must be stricken from the case. Nine articles remained in the suit, set for trial by jury January 3, 1968, in Upper Marlboro.[39]

The trial consisted of five days of testimony and arguments heard before the circuit judge Robert B. Mathias. Chasanow charged that the paper had made his client "the most hated man in Greenbelt through a series of false and malicious articles."[40] Chasanow based his case on the claim that Alfred Skolnik and the paper had led a communitywide effort to discredit Bresler in Greenbelt and that "they had knowingly and with malice published false information about the plaintiff in a deliberate attempt to damage his reputation." Chasanow charged there was a strong link between Skolnik, the paper, Citizens for a Planned Greenbelt, and Greenbelt Homes, as well as city council members and others in the community. Part of CFPG's money from a fund-raising drive had been used for a lawsuit in which a number of town residents sued Bresler for damages, charging that he had not fulfilled his responsibilities toward them. Chasanow implied that these actions were part of a conspiracy against Bresler. Finally, Bresler felt that his reputation had been damaged by those who accused him of "blackmail."[41]

The defense lawyers Roger A. Clark and David Reich charged that Bresler's suit was an effort to "stifle legitimate criticism of his public dealings by civic-minded citizens bringing news to the community."[42] They argued that statements regarding Bresler in the paper represented "fair comment about his public dealings as a land developer in Greenbelt." As a public figure, newspaper comments about him were within the constitutional guarantee of freedom of the press. Regarding the use of the term *blackmail*, Clark noted that the community's concern about Bresler's activities and expressions of this concern, including the use of the term *blackmail*, did not imply any criminal charge but had simply been made at council meetings and faithfully reported by the paper.[43]

The jury deliberated for three hours before reaching a verdict of guilty against Alfred Skolnik and the *News Review*. It awarded Bresler $5,000 in compensatory damages and $12,500 in punitive damages. The response in town only took as long as the next issue of the paper, which contained a boldface front-page advertisement inserted by the Freedom of the Press Committee. It began, "A Call to the People of Greenbelt." The advertisement observed that, "in a very real sense, the community has been on trial along with the *News Review* and Mr. Alfred Skolnik, president of the Greenbelt Cooperative Publishing Association." The committee invited residents to a meeting the next evening to "discuss necessary proposals and steps to help carry the case through an appeal of the decision." The advertisement ended, in boldface, "Come to the Defense of Freedom of the Press."[44]

On January 12 more than 130 people gathered at the municipal building to listen as a committee member pronounced: "Seldom in our lifetimes are we privileged to make a direct contribution to protect one of our basic freedoms—freedom of the press. This is not just the *News Review*'s right to report the news, but your right to be kept fully informed on what is going on in the community." The group voted unanimously to support the *News Review* in an appeal of the libel verdict. They also voted to set up an escrow fund of voluntary contributions, to serve as security for the $17,500 judgment while the case remained under appeal. The supportive crowd pledged almost $12,000 at the meeting. The Freedom of the Press Committee also hoped to

National Aeronautics and Space Administration
(immediately east of Greenbelt. Photo courtesy
*Greenbelt News Review*

collect an additional $10,000 for future legal fees. A fund-raising committee began work, considering the possibilities of a dinner or dance rally, door-to-door solicitation, or a fun night.[45] More than 2,000 residents had already contributed over $5,000 for the original trial.[46]

The Freedom of the Press Committee continued its publicity in the January 18 issue of the paper, where a large advertisement on the second page asked "ARE YOU AMONG THOSE PRESENT?" urging readers to make a contribution and to attend the mass meeting set for January 26. At this meeting, Alfred Skolnik reported that he had received an offer from Chasanow, on Bresler's behalf, indicating that he would be willing to forgo the judgment if the paper paid $5,500 in legal fees and $1,000 in court costs. The one hundred people present unanimously urged Skolnik to appeal the decision. The committee planned a "gala dance affair" for May 11 to raise funds for the appeal.[47]

As the paper's board of directors made the decision to appeal, they observed that "the case involves a basic constitutional question which merits review at the highest court levels, including the U.S. Supreme Court if necessary."[48] Weekly advertisements in the paper kept residents informed of the amount contributed and the amount still needed for the paper's appeal. By

February 15, $15,700 had been raised. In March, front-page weekly cartoons by Isidore Parker began appearing, pointing out the importance of the *News Review* ball, which raised sufficient money for the appeal.[49]

Even while raising money for the newspaper suit defense, Greenbelters continued to monitor development issues. In January 1968 the city council approved two resolutions permitting the housing authority of Prince George's County to operate a leased housing program for low-income families within Greenbelt city limits. At an April meeting, Citizens for a Planned Greenbelt listened to a guest speaker give tips on fighting zoning applications that were counter to the community interest. The speaker advised "careful case preparations for zoning hearings, employment of a competent attorney who has worked with citizens' groups, and use of witnesses with credentials in the field of planning, land engineering, real estate and construction."[50] Greenbelters used this advice, not to fight against low-income housing, as would have been true in many other places, but to fight against housing that did not adhere to the town's original planned community concepts.

In August the MNCPPC released its long-awaited and much revised "Area 13-N Master Plan" for the College Park–Greenbelt area and presented it to the county commissioners. The plan no longer contained four-lane highways through the center of Greenbelt or high-rise apartment buildings in the Center. The municipalities concerned had thirty days to make their opinions of the plan known. At a special city council meeting, the council expressed its general approval, saying it was "a great improvement over earlier versions," which the county commissioners had rejected in July 1965 because they felt the proposed development had been at too high a density. The new master plan contained zoning for single-family homes for most undeveloped parcels. An editorial in the *News Review* expressed the feeling of Greenbelters: "The revised Area 13 plan is a tribute both to the citizenry of Greenbelt and the MNCPPC—to the citizenry because of its unflinching determination to make its views known, utilizing its legitimate rights of petition, assembly, and free speech; to the MNCPPC for its recognition that the residents of a city and its elected representatives have a paramount interest in any zoning and development plans that affect their community."[51]

Even though the new master plan appeared favorable, Greenbelt residents maintained their vigilance. In November, Citizens for a Planned Greenbelt, as well as the city council, went on record as opposing Charles Bresler's request to build more apartments in his Lakeside North development. The Prince

George's County planning board rejected plans by Sidney Brown to add to Beltway Plaza until he actually completed a road he had promised to build.

In addition to opposing unwanted development, Greenbelt residents continued to work actively for the kind of community they wanted. On November 26 residents agreed to a $950,000 bond issue to finance several projects: an addition to the youth center, construction of a recreation building in Springhill Lake, acquisition of parcels 7 and 8 near Boxwood Village for parkland, construction of tennis courts and other park facilities, and roadway improvements. After the passage of the bond issue, the paper editorialized on the wisdom of Greenbelt residents who have "a quality of farsightedness that has continually marked the community's history. In the greater interest of the entire community, a majority of Greenbelters have voluntarily chosen to tax themselves to provide a better place to live in."[52]

In December the circuit court judge, Perry Bowen, ordered Bresler to convey 3.3 acres of the Charlestowne tract to the city of Greenbelt, as promised in an agreement signed by Bresler in April 1962. The judge stated that the city had lived up to its part of the agreement but that Bresler had not. Bresler contended that the city failed to bring timely action on the dispute. The judge commented that "Bresler's argument regarding the delay in filing was based on an assumption that the city should have anticipated that he would have violated his promises."[53] In March 1969 Judge Robert Mathias decided the last disputes between the city and Bresler, enjoining Bresler from building any additional units in Charlestowne Village. Bresler agreed to drop his appeal of the order to convey 3.3 acres to the city.

Thus by 1969 the tide had clearly turned in favor of residents and against massive development. On March 12 the Maryland–National Capital Park and Planning Commission officially adopted the College Park–Greenbelt and Vicinity Master Plan in a form that basically agreed with the city's wishes. On November 27 the county commissioners unanimously adopted the plan, in which the estimated population of Greenbelt would grow to 40,000, with 13,500 dwelling units, rather than 50,000 as in the original plan. The advantage of the new MNCPPC plan became apparent when Charles Bresler and Theodore Lerner asked for apartment zoning for parcels 1 and 2. The zoning commission denied their requests because they conflicted with the master plan's zoning of single-family homes. It is important to note that by 1969, when these decisions in accord with Greenbelters wishes were being made, much change had already occurred. Greenbelt's population of 7,479 in 1960 had risen to 18,199 in 1970. The municipal budget had tripled, but the enormous increase in taxable property allowed the tax rate to be halved.[54]

The context in which Greenbelters struggled against development changed in the later part of the 1960s, aiding them in their fight. Although the first to resist the efforts of speculators, others joined their ranks, led by Gladys Spellman, a Prince George's County commissioner who became a U.S. congressional representative at this time. Prince George's County citizens, disgusted with their local leaders and desirous of a better form of government, passed a referendum in 1970 creating an eleven-member county council and an elected county executive, who promptly adopted much stricter planning and zoning procedures. The Maryland historian George Callcott said of this period: "By the late 60's suburban voters had generally prevailed over their politicians, and planning, zoning, open hearings, and reasoned exceptions to the plans were all an accepted part of suburban life."[55]

After many delays, justice was served through a series of trials in the early 1970s. Jesse Baggett, the county commissioner, was convicted of accepting bribes from developers and sentenced to serve fifteen months in prison and pay a $5,000 fine. The chairman of the planning board, William Stevens, received a sentence of nine months in prison for income tax evasion; the banker William Kahler was convicted of perjury; local developer Ralph Rocks was sentenced for bribing county officials.[56]

As the 1960s ended, the libel suit by developer Charles Bresler against the *News Review* continued to claim Greenbelters attention, time, and energy. The Maryland Court of Appeals heard oral arguments in the case on February 4, 1969. On May 2, the court announced its unanimous decision to affirm the $17,500 libel judgment awarded to Bresler against the *News Review*. The court rejected the appellant's argument that the articles in question were "constitutionally protected because they were accurate reports of what was said during official public meetings of the Greenbelt city council regarding a public proposal of great interest to the community." The ruling stated that the federal constitutional protection provided by the case of the *New York Times* did not apply, as there was "credible evidence from which the jury could find that the newspaper published the articles with actual knowledge of the falsity of the charge of the commission of the crime of blackmail." (The *New York Times* rule "precludes recovery in libel suits by public officials and public figures except where they can prove that a false and defamatory statement was made with knowledge that it was false or with reckless disregard of whether it was false or not.") The presiding judge stated that "nothing at the public meeting justified the charge that Bresler had committed blackmail."[57]

This decision caused great disappointment in Greenbelt, as had the initial one. Albert Herling and Charles Schwan, co-chairmen of the Greenbelt Freedom of the Press Committee, sent a letter to the editor "to express continuing complete confidence in and full support for the *News Review*."[58] In an interview regarding his continuing work on behalf of the paper, Al Herling explained: "If something is good, it's worth fighting for. I'm somewhat utopian. You do what you have to do."[59]

The newspaper staff decided to appeal the libel case to the U.S. Supreme Court, filing a writ of certiorari in August 1969. In October the court agreed to hear the case. Briefs were filed in December, and oral arguments began on February 24, 1970. On May 18 the court announced its unanimous decision to reverse the $17,500 libel judgment against the paper. The court decided in favor of the paper on two minor issues regarding instructions given to the jury in the original trial and on the question of whether the paper showed "malice" in publishing its accounts. Mr. Justice Stewart addressed the major point as follows:

> The *Greenbelt News Review* was performing its wholly legitimate function as a community newspaper when it published full reports of these public debates in its news columns. If the reports had been truncated or distorted in such a way as to extract the word "blackmail" from the context in which it was used at the public meeting, this would be a different case. But the reports were accurate and full. . . . It is simply impossible to believe that a reader who reached the word "blackmail" in either article would not have understood exactly what was meant: it was Bresler's public and wholly legal negotiating proposals that were being criticized. No reader could have thought that either the speakers at the meetings or the newspaper articles reporting their words were charging Bresler with the commission of a criminal offense. . . . Indeed, the record is completely devoid of evidence that anyone in the city of Greenbelt or anywhere else thought Bresler had been charged with a crime.
>
> To permit the infliction of financial liability upon the petitioners for publishing these two news articles would subvert the most fundamental meaning of a free press, protected by the First and Fourteenth Amendments. Accordingly, we reverse the judgment and remand the case to the Court of Appeals of Maryland for further proceedings not inconsistent with this opinion.[60]

On July 7 the Maryland Court of Appeals issued an order reversing and vacating the libel judgment and required Bresler to pay the newspaper the court costs incurred in defending the action.[61]

The free speech issues contained in the *News Review* case were significant enough for widespread notice to be taken of the result. This began at home, where the mayor and the city council passed a resolution "congratulating the *Greenbelt News Review*, its staff and Alfred M. Skolnik, president of the Greenbelt Cooperative Publishing Company on their successful defense of the right to a free press." The *Washington Post* provided its readers with all the details of the case, as it involved local events. Other accounts focused on the issue of freedom of the press such as the *New York Times* article, beginning: "The Supreme Court broadened the constitutional defenses of newsmen against libel suits today by ruling that the press cannot be held liable for reporting exaggerated charges leveled against public figures when it is clear that the accusations are 'political hyperbole.'"[62]

The *Washington Evening Star* interpreted the case this way: "Newspapers won clearance from the Supreme Court today to report in full controversial public meetings even if 'vigorous epithets' are used against a public figure. . . . The high court said the First Amendment protects news articles even if hot words are used against an individual." The Associated Press release commented: "The decision assured the press and broadcasters they cannot be held for libel for reporting free-swinging debate on public issues. . . . The decision went on to say that the First Amendment strictly limits the scope of libel judgments so that the press may freely report issues of community concern without financial worry."[63] In 1987, at the *News Review*'s fiftieth anniversary dinner, Roger Clark, who represented the paper in the court suit, announced that in the intervening seventeen years more than 250 court decisions had cited the Greenbelt case as precedent.[64]

An announcement in the August 27, 1970, *News Review* informed residents that the "press escrow fund" was "available for refunds." The $20,000 in Twin Pines Savings and Loan would be returned, with interest, never having been needed. The co-chairmen of the Freedom of the Press Committee, on its dissolution, expressed appreciation to the community for its support, acknowledging the debt owed Roger Clark, who "successfully carried the case from the trial court to the U.S. Supreme Court on a volunteer basis."

The *News Review* served as a logical target for Charles Bresler during the period of his development activities in Greenbelt. He correctly perceived that if the paper had not existed, he probably would have had his way in the town. The residents of Greenbelt could not have banded together to fight against his plans for the community if they had not been kept informed of current events. The *News Review* not only told Greenbelters of Bresler's activities, it

Maryland Trade Center, intersection of Capital Beltway and
Baltimore-Washington Parkway. Photo by Blue Ridge Aerial
Surveys, courtesy Greenbelt Museum

also enabled them to communicate with each other to form organized opposition. Residents banded together and eventually prevailed against developers
who threatened the concept of Greenbelt as a planned cooperative community. The attacks on Greenbelters from the outside can be seen as fortunate in
one way, as they caused townspeople to unite. This centrifugal force counter-
balanced serious divisions within the community, which are addressed in the
next chapter.

The havoc that developers wreaked on the landscape with their land spec-
ulation affected much more than housing, as the county lacked the infra-
structure to meet the needs of its residents. The overcrowding of schools
became a constant worry to Greenbelt parents by the mid-1960s. In September 1967 work began on a two-story addition to Center School, containing
eight classrooms and a library. The county board of education scheduled
North End Elementary School for an addition, as it had experienced a 34 percent increase in pupils in 1967 alone. Springhill Lake had its own elementary
school, so Greenbelt now had three. The 1968–69 school year saw the open-

ing of newly built Parkdale Senior High School, located on Good Luck Road south of Greenbelt Park, and an addition to Greenbelt Junior High School. However, in the fall of 1969, in its second year of operation, Parkdale High had 2,700 students in a school built for 1,800.

Greenbelters desired not only good schools for their children but also schools consistent with the goals and design of the community as a whole. In a lengthy letter published in the August 20, 1970, *News Review,* a resident discussed the history of such concern in Greenbelt. The author reviewed the cooperative nursery school formed in the first year, a cooperative kindergarten begun after the federal government withdrew its funding, the PTA's formation of a foreign-language program in the elementary schools, later incorporated into the regular school program, and special after-school and weekend programs initiated by the community. The letter concluded:

> When the question of a new high school came up, education-oriented Greenbelt got together and chose a site (not parcel 2) that would preserve the original lay-out of the City as a planned, safe, walk-and-bicycle community while at the same time offering the best possible educational facilities. The city backed up its choice by offering 10 acres of land, while the GHI [Greenbelt Homes] offered 10 acres at the mortgage-release price. For four dark years the realization seemed far off. Now we start to see the light at the end of the tunnel. . . . During the recent discussions the word "Utopia" has been used. Greenbelt is as close to Utopia as one can get in real life. However, a practical, down-to-earth Utopia requires perseverance, patience, sometimes quite heated debates and above all: long and hard work.

The people of Greenbelt proved themselves capable of doing this "long and hard work" to keep their town as close as possible to their vision of an ideal community. Even while struggling against developers, they fought a four-year battle with the Prince George's County Board of Education to obtain schools that would fulfill community requirements.

To meet the urgent need for more schools, the board of education decided to build a complex including senior, junior, and elementary schools on parcel 2, located between Greenbelt Homes property and the Baltimore-Washington Parkway. This land had been a source of dispute between Bresler and Greenbelt residents when he wished to have higher-density zoning in exchange for selling part of it for school sites. Residents remained firmly against use of this site for a school, as it had access problems, bounded as it was by a residential neighborhood and a superhighway. A four-lane street would have to be put through Greenbelt Homes property to provide sufficient roadway. Greenbelt Homes filed suit against the school board to stop the location of the school in parcel 2, but this ploy failed, as the courts ruled in favor of the school board. Citizens collected petitions, held mass meetings, and wrote to the governor, the state superintendent of schools, state legislators, and county commissioners to no avail. The school board remained supremely indifferent to the feelings of the community. Out of this turmoil, predictably, a new Greenbelt citizens group was formed: the Save Our Community Committee (SOCC).[65]

For a brief period the city council and the mayor agreed with the plan recommended by the board of education, but after 1,100 residents signed petitions against the site, they changed their minds. In April 1970 city officials and SOCC members went to Annapolis and managed to get a resolution passed instructing the school board to explore alternative sites. An open letter from the SOCC vice chairman to Governor Marvin Mandel asking him to sign the resolution, published in the April 16 News Review, revealed the ability of Greenbelters to use politics to their advantage: "Governor Mandel, on behalf of a city united to protect its original historically unique land-use plan, and on behalf of the many disillusioned citizens here who are growing convinced that the only remaining avenue open to them is confrontation politics, I appeal to your good political common sense in this election year."

In June Greenbelters went to the county commissioners asking that money for the school be delayed for a year. In August the commissioners agreed and received the thanks of the community by being invited to the annual fish fry held at the lake. In September a newly elected school superintendent and school board announced that a hearing regarding the site of the proposed high school would be held and that the new school would be called Franklin Roosevelt High School to honor the contributions the former president made to Greenbelt. On September 17 more than 300 Greenbelters attended the public hearing to represent the views of the community. In October the board agreed that a different site would be selected for the school. It is interesting to note that citizens of Prince George's County elected a new and more receptive school board in 1970 at the same time that a new and comparatively innovative county government began work.

In February 1971 the board of education authorized the purchase of forty acres for Franklin Roosevelt High School located in the triangle between the Baltimore-Washington Parkway and Greenbelt Road. Construction began in November 1973. In December 1975 several members of the board of education suddenly announced that half of the school should become a technical or trade school. More than thirty-five Greenbelters appeared at a school board meeting in December to discuss various proposals for the school, among them a magnet program in math and science. Residents voiced their concern about the tardiness of these proposals, as the school's opening date was set for the fall of 1976. The Greenbelt Citizens Committee for FDR High School materialized to monitor the situation.

In February, after much argument, the school board decided that half the school would be a comprehensive high school, including all Greenbelt students and those from nearby areas, and the other half would be decided by school staff. In March the staff announced that the school would house a technology center offering advanced courses in math and science. In addition, the building would have special facilities for the performing arts, such as a well-equipped stage and television studio. School capacity would be 2,300, with about 1,500 expected the first year.

As if last-minute controversy about the use of the school was not enough, in December 1975 a debate began regarding the name of the school. A letter to the editor from a former Greenbelt resident, Margaret Wolfe, appeared in the *Washington Star,* reminding readers that "about ten years ago, when I was a resident of Greenbelt, Md., proposals were made to name the new Greenbelt high school for Eleanor Roosevelt." The *Star* published a picture of Eleanor as a young mother above the letter and headed it "Greenbelt Should Honor Eleanor." In response, Edna Benefiel, another former resident, wrote to the *Star* agreeing with Margaret Wolfe. She remembered Mrs. Roosevelt: "She was often a visitor to the community, talking with pupils at school and driving around the town. . . . If the present Board of Education won't review past history and honor one of the greatest helpers Greenbelt ever had, a woman who understood hardworking people, by naming the new high school for her, then I feel that ghosts will linger in that lovely building every night."[66]

The actions of Margaret Wolfe and Edna Benefiel show that at least some of those who left Greenbelt behind remained interested in its community affairs. While those in Greenbelt worked to achieve the type of school they wished for their children, past residents had the leisure to be concerned with the name of the school and cared enough to take action.

Several days after Edna Benefiel's letter in the *Star,* the January 8, 1976, *Prince George's Post* editorialized, "Naming the new senior high school for Eleanor Roosevelt would honor a great humanitarian woman who richly deserves such credit." The following week the school board voted to name the new school Eleanor Roosevelt High School. Forty years after she watched the new town come into being, Greenbelt remembered Eleanor Roosevelt as a key player in its struggle for completion as a planned community.

One other structure rising in the community other than the new high school had the full blessing of residents. Since 1963 Greenbelters had been actively working to get a new library in their town. The Greenbelt Library began in 1939 as a city-supported venture, begun by Greenbelt "pioneers" in Center School. However, it lacked sufficient space in its location in two classrooms in the school. A group of Greenbelt residents consistently attended public meetings, vociferously stating the need for a new library, until the county commissioners appropriated necessary funds in the 1966 county budget. The county board of education made land that it owned adjacent to Center School available for the library, exactly where Greenbelters wanted it, as it fit perfectly into the original plan. The library would be next to the community center, adjacent to the shopping center, and convenient for pedestrians.

More than 120 Greenbelters and local dignitaries attended groundbreaking ceremonies on March 10, 1968. Al Herling introduced members of the Greenbelt Library Association, noting, "I think we have the readingest population not only in the county but in the state." The Greenbelt Library did have the highest per capita circulation in the county. Rabbi Weisenberg of Mishkan Torah delivered the invocation and the Reverend Birner of Holy Lutheran Church gave the benediction. A reception in the city council chambers featured an exhibit of Greenbelt memorabilia, including Farm Security Administration photographs borrowed from the Library of Congress.[67]

The town dedicated its new library on April 7, 1970. In his speech, Al Herling summarized the effort that had been necessary to achieve this goal, including the night in January 1965 when the group from Greenbelt waited until 2:00 A.M. to testify regarding the need for a new library. He pointed out that an important part of the collection would be material on the history of Greenbelt, community planning, and consumer cooperatives, to be housed in the Rexford Guy Tugwell room. Since 1968 both the Friends of the Greenbelt Library and the Greenbelt Historical Society had been gathering materials for inclusion in this special collection. Herling reported that he had informed Tugwell, who lived in Santa Barbara, California, of the collection and the room named in his honor: "I must report to you that Professor Tugwell was deeply touched and extremely gratified that his contribution to the life of the people of our country in general and Greenbelt in particular is being recognized." Herling also described Greenbelt in April 1970:

> There is a spirit about Greenbelt which time has not erased. There is a sense of community that surely must create love on the part of many. There is still always controversy in Greenbelt; not all of it reasoned; some of it bitter; but on the whole, people in this community have learned to disagree without being disagreeable in the process. When a community goal seems hopeless, when even the efforts of our elected representatives to achieve what the community desires seem to have met a dead end, there are always those spirits in the community who, joining together, issue a clarion call for action which renews hope and which once again stirs the community to united endeavor."[68]

At the same time the new library appeared, the Woman's Club of Greenbelt achieved its goal of memorializing Eleanor Roosevelt. The club put an ornamental iron fence around a hundred-year-old tree at the spot where Eleanor Roosevelt had stood to survey the ongoing construction of Greenbelt. On the fence they placed a plaque with the following:

> Anna Eleanor Roosevelt, 1884–1962. First Chairman of the U.N. Commission on Human Rights.
>    From this vantage point she surveyed the site and spurred the work from which sprang Greenbelt, the first planned garden community in the land dedicated to the unfolding and uplift of the human spirit and light, air and the greenery of open space. Through her a blow was struck for the long-suffering and a beacon lit for America.

The dedication of the Eleanor Roosevelt Tree occurred on November 2, 1968, with the plaque unveiled by Mrs. Roosevelt's daughter, Mrs. James A. Halsted.[69]

Even while Greenbelt remembered its past, the outside world increasingly remembered Greenbelt. The tremendous amount and pace of development in the decade of the 1960s, as well as a new interest in planning in the form of new towns such as Reston, Virginia, and Columbia, Maryland, stimulated renewed interest in Greenbelt as a planned community. The architect Albert Mayer visited all three "green towns" in the summer of 1966 to write *Greenbelt Towns Revisited*. His analysis focused on town design in order to provide useful information for planners active in the new town movement.

The *Washington Post* published an editorial on April 15, 1964, titled "Lament for Greenbelt": "Sadly betrayed by the Federal and local government, the town of Greenbelt today is hardly more than a well-planned subdivision

Intersection of Capital Beltway and Kenilworth Avenue
(Capital Office Park under construction north of the
Beltway, Springhill Lake to the south). Photo by
Commercial Photographers, courtesy Greenbelt Museum

surrounded by large apartments. . . . A noble conception is being destroyed by official negligence." A year later the August 4 *Washington Star* editorialized:

> There are too few suburban communities that can claim a pleasingly distinct character. Greenbelt is one of them. To say that Greenbelt is worth protecting has little to do with sentimentalism over the New Deal experiment which created the city. It has everything to do with the fact that certain amenities built into the community have more validity than ever before as suburban pressures mount. Chief among them is the provision of low-cost housing in a low-density, pleasant environment.

The architectural critic Wolf Von Eckardt wrote a lengthy account titled "Greenbelt—A Town's Fading Dream," which appeared in the Sunday *Washington Post* magazine on July 23, 1967. The largely negative article prompted defense by residents in the form of letters to the editor, just as residents had defended the town from attack a generation before. The *Baltimore News American* published a lengthy history titled "Greenbelt: A Planned City" on

August 17, 1969. Neutral in tone, the article did not evoke protest as Von Eckardt's did.

Most outside attention on Greenbelt in the 1960s focused on the actions of developers and their effect on Greenbelt. A *Saturday Evening Post* article, "The Rape of the Land," featured Greenbelt as one of its examples. A *Washingtonian* magazine article, "Greenbelt Has Been Deflowered, But You Can't Cry Rape," disparaged the town, past and present. A typical comment regarding the residents: "They are all sadly aware that they can never recapture the dream of their town: it is being buried under the new developments that have sprung up in the last six years."[70]

Not a single author of these newspaper and journal articles wrote about, or apparently even noticed, the vigorous actions people in Greenbelt took regarding their destiny. Only a local paper, the *Prince George's Sentinel*, in its July 18, 1968, issue responded to the *Washingtonian* article with perception:

> Greenbelt residents are understandably angry about a recent article in a Washington magazine that depicts their town, in the irate words of Councilman Francis White, as "a run-down, left-over WPA community."
>
> It is inconceivable that the freelance author of the article, if he had done any research at all, could have overlooked the real Greenbelt in his fascination for emphasizing its aging row houses and abortive deals with developers.
>
> A significant point he ignored is that these row houses, which were built during the days of the New Deal, are still coveted as dwelling places by many who could afford much more so-called luxurious housing.
>
> Perhaps the main reason for this is that the City of Greenbelt is not just an accumulation of houses.
>
> Its City Council meetings, for example, are the best models of grassroots democracy in action found outside the New England town halls. Every citizen who so desires is given a chance to have his say.
>
> The city's park and recreation facilities are tops in this area.
>
> And, most important, there is the intangible benefit of community spirit, of having a common purpose with one's neighbors.
>
> To many residents of suburbia, these true luxuries are more valuable than those found in new subdivisions or swank high-rise apartments.
>
> Let those who read the critical magazine article be assured that the Greenbelt dream is not dead. If anything, it is more alive now than it was 30 years ago.

Change continued in Greenbelt during the 1970s but more slowly than during the previous decade. A comparison of 1970 and 1980 census figures shows that the city actually lost population during the decade, decreasing from 18,199 residents in 1970 to 17,332 in 1980. (See appendix.) However, the number of blacks increased sevenfold. Even though the number of residents declined, the number of housing units grew from 6,519 in 1970 to 8,005 in 1980. Greenbelt thus reflected the national trend of decreasing household size. The number of nonfamily households increased by 160 percent, comprising 44 percent of the population in 1980. The growing number of both older individuals living alone and University of Maryland students residing in Springhill Lake caused most of this change. The percentage of rental units decreased from 68 percent in 1970 to 60 percent in 1980, as most new housing took the form of condominiums or town houses in Greenbelt East, east of the Baltimore-Washington Parkway. During this decade the number of residents living in areas separated from the Center by freeways greatly increased. Led by the development of Greenbelt East, Greenbelt became an entity composed of distinct parts, making it difficult to create a feeling of community in the outlying areas.[71]

During the 1970s and into the 1980s, the city council objected to all new development because roads in the area were used well beyond capacity. Developers ignored these protests, as Greenbelt held much potential for them, with 65 percent of the land in town still undeveloped as of 1970.[72] Because of reforms in both county law and government, the master plan remained the official guideline unless a developer managed to appeal successfully. A state moratorium on new sewer hookups from 1970 to 1977 also affected growth.

The halt on sewer construction caused difficulties in Greenbelt for the developer Alan Kay, who wished to build his Greenbriar project in the triangle between the Baltimore-Washington Parkway, Greenbelt Road, and the Goddard Space Flight Center. Kay offered to build his own pumping station and sewage line to serve Greenbriar, but the county planning board vetoed the plan, saying enlargement of the existing sewage plant had to occur first. Kay also applied for a zoning exemption to build more bedrooms than would normally be allowed in his "luxury-type garden apartments." In August 1971 the city council agreed to allow zoning exemptions if the builder repaid the city for purchasing five acres to be used for recreation and open space. The next week, on August 19, the following letter appeared in the *News Review*:

> Dear Mayor, Council Members and Citizens of Greenbelt:
>     I want to take this opportunity to thank the Mayor and City Council for arriving at the positive decision concerning the proposed Greenbriar Apartment project in Greenbelt, Maryland. I know this decision was reached only after many trying hours of deliberation and soul searching in evaluating what is in the best interests of the City of Greenbelt.

The citizens of Greenbelt are also to be commended for taking such a keen interest in developments that affect their City.

I believe with the accord that has been achieved between Greenbelt and the developer, a new era of mutual concern and accommodation has begun, and we shall do all that is possible to enlarge upon and encourage this atmosphere.

Again, many thanks to the government and citizens of Greenbelt.

Very truly yours,<br>Alan I. Kay

This letter reflected the new context for development. Kay surely was aware of Bresler's bruising battles with the city and decided that a conciliatory approach would be more productive.

In November 1971, when Kay applied for permits to build a temporary sewage treatment plant for Greenbriar until a permanent one would be possible, more than forty people attended a three-hour meeting held in response to his plan, where they heard the testimony of state experts from the Maryland State Department of Water Resources and the Prince George's County Department of Health.

When the Prince George's County Council approved the exemption allowing Greenbriar to have more units per acre than zoning allowed, the Save Our Community Committee brought suit in circuit court. The court rejected the suit, but SOCC appealed the decision. However, the group eventually dropped its appeal due to lack of funds. This failure demonstrates the less intense interest by Greenbelters in development occurring outside the central core of town than in developments inside the freeways.

In July 1973 the Washington Suburban Sanitary Commission granted a permit for construction of an on-site temporary sewage plant at Greenbriar. However, the National Aeronautics and Space Administration turned down a request for a sewer line to run through its property. In September Kay again turned to the *News Review,* publishing an "Open Letter to the Citizens of Greenbelt" explaining how safe and excellent the treatment plant at Greenbriar would be. A week later another letter from Kay spoke of the tax benefits to the city if Greenbriar could be built. By these actions, Kay became the first developer to utilize the paper to join in the community dialogue.

In late September Kay filed suit against the state secretary of health and mental hygiene and the deputy state health officer, asking the Circuit Court of Prince George's County to grant a mandatory injunction requiring them to issue a construction permit for the temporary sewage treatment plant. According to stories in the *Washington Post* and the *Baltimore Sun,* Kay tried

Kenilworth Avenue (Greenbelt Road crossing to south),
June 1987. Photo courtesy *Greenbelt News Review*

to get Governor Mandel and Lieutenant Governor Blair Lee to intercede with
state officials on his behalf.[73] Construction at Greenbriar began to move
ahead in February 1974, when Kay finally received state and county council
permits for the on-site treatment center. He also wished to build an extra
building, which would put him in violation of the green space requirement
mandating land to be reserved for parks. At the Prince George's County
Board of Appeals four Greenbelt councilmen and the city manager, Jim
Giese, testified against the additional development. NASA agreed to let Kay
run a sewer line through its property, and the appeals board granted a vari-
ance so Kay could build the sewage plant and an extra apartment building.[74]

On April 25, 1974, Kay ran full-page advertisements in the *News Review*:
"Greenbriar Associates cordially invites the residents of Greenbelt to a preview showing of the Greenbriar condominium luxury apartments." The next
week Kay appeared before a special session of the city council, promising to
adhere to all agreements he had previously made with the city. In September
the U.S. Environmental Protection Agency gave a discharge permit to the
Greenbriar sewage plant, followed in October by a permit from the Washington Suburban Sanitary Commission. This final permit allowed occupancy
of the first building. Kay originally planned Greenbriar to have 1,193 condominiums in nine buildings. However, due to a slowdown in the real estate
market, some of the last buildings emerged as rental units instead.

The campaign of SOCC and its leader, Rhea Cohen, slowed down the
development of Greenbriar but did not stop it. The community at large felt
sympathy for Cohen's activities but did not become intensely involved. Because of Greenbriar's location across the Baltimore-Washington Parkway, it

did not seem as crucial as it would have had Greenbriar been built next to the lake or elsewhere in the central part of town.

This does not mean that residents maintained no interest in development on the periphery. The developer Sidney Brown's actions at Beltway Plaza provided a continuing source of irritation. In late 1971 the Prince George's County Board of Licensing and Permits would not allow an additional twenty-three stores to open until Brown completed his promised grading and greening of the slopes north of the plaza. The December 2 *News Review* applauded this move in an editorial entitled "A Refreshing Change." Brown did not appreciate the paper's coverage of issues regarding Beltway Plaza. In a scathing letter to the editor on January 13, 1972, he charged the paper with ignoring the important story of the largest shopping center in Prince George's County being developed right in their town in favor of focusing on "a small erosion and pollution problem." This of course prompted a response from a Greenbelt family via a letter to the editor on January 20:

> We never drive past the clay mountain to the right of S. Klein's without thinking of the two men who lost their lives there during initial construction of Greenbelt's Shopping Mall. How does one take the measure of a man who has left the landscape around Klein's defaced for the past ten years, is now scheduled for trial with the state of Maryland for alleged pollution and dismisses the whole issue lightly as "an anti-erosion kick by a minute segment of the community?" Mr. Sidney J. Brown must come to realize that the day is long past when Greenbelt citizens will sacrifice health, safety, and natural resources for a little extra shopping convenience and increased tax base.

Even though Beltway Plaza is across the Beltway from the central area of town, Greenbelters felt strongly about its development, especially its negative effects on the environment.

Further commercial development began in 1974 with plans for the Golden Triangle, a fifty-seven-acre tract formed by Kenilworth Avenue, the Beltway, and Greenbelt Road. The city wanted right of approval over final site development plans, but the county planning board denied this. Capitol Cadillac initiated development of the first tract, while Prudential Insurance bought much of the remaining land. In October 1977 when work finally began in the Golden Triangle, the developer cleared half the land of trees, including those that had been flagged for saving. City council members protested at a county executives' meeting but were ignored. Greenbelt residents continued to be incensed by any decrease of green, especially such wholesale removal of trees, keeping up a resistance that remained unique among Prince George's County communities. City protests had no effect on slowing down the building of Windsor Green, a townhouse development south of Greenbelt Road opposite the new high school.

The owners of Springhill Lake began new commercial ventures in April 1978 when they received all the necessary permits. They announced plans to build five ten-story office buildings and a 200-room motor inn with convention center and restaurant on their thirty-two acres across the Beltway from Springhill Lake. The development of this area, known as Capital Office Park, heralded the beginning of an office construction boom in Greenbelt. Most of the construction in the 1980s continued to be office parks.

Greenbelt now contains three office parks: Capital Office Park, Hanover Office Park, and Maryland Trade Center Buildings, the latter in Greenbelt East, adjacent to the newly built Greenway Shopping Center. This type of development pleased the city manager, James Giese, because occupants of office buildings do not place as many demands on city services as residents do, while the buildings increase the tax base. By 1986 the daytime, office worker population of Greenbelt was estimated to be similar to its resident population of approximately 18,000. Thus fifty years after its creation, Greenbelt provided a place to both live and work, as called for in the British garden city concept.[75]

One particular building in Capital Office Park, opening for business on November 14, 1985, demonstrated Greenbelt's new image: the Greenbelt Hilton. It seemed to fit in well with Capitol Cadillac in the Golden Triangle. A man who moved into Greenbelt as an eight-year-old in 1938 and who is still in Greenbelt said: "The single most momentous thing was to see a Greenbelt Hilton go up. Never, never in my wildest dreams did I think that my wife and I would go over to have dinner at the Greenbelt Hilton. Never occurred to me that would happen."[76] The fiftieth anniversary dinner dance, held at the Hilton, provided visiting Greenbelters with a first-hand impression of changes in town.

The area where the Baltimore-Washington Parkway, the Beltway, Greenbelt Road, and Kenilworth Avenue come together is called the "billion-dollar circle" by county officials. The location is "one of the hottest real estate investment and development territories on the East Coast."[77] An article in the *Prince George's Journal* found the cause in the "high technology lures" of Goddard Space Flight Center, the Beltsville Agricultural Research Station, and the University of Maryland, as well as the network of freeways providing access.[78] The office building boom caused an increase in traffic far beyond road capacity, in turn causing much stress on local residents. Greenbelters protested in vain about overcrowded roads, as the county did not even begin to catch up with development until 1986 with the widening of Greenbelt Road and the completion of its interchange with Kenilworth Avenue. This ever-widening roadway system illustrates the dramatic changes that development wrought on the city.

Throughout the 1980s and 1990s, residents of Greenbelt and the city administration continued in their efforts not only to thwart unwanted development but also to actively pursue growth and change that would be in accordance with local desires. As a *News Review* editorial of February 4, 1971, commented, "Up to now we have been fighting a rearguard action, securing postponements in zoning decisions, and purchasing certain crucial but small parcels of land, which, if developed, would have presented the most flagrant abuse of the 'planned community' concept. Now is the time for us to strike out boldly and imaginatively." The editorial suggested the purchase by the city of parcels 1 and 2, part of the remaining greenbelt. This article reflected the belief that it was not only possible but even timely for Greenbelt to change from a defensive position to an offense that actively reclaimed what remained of the original greenbelt. In January the city purchased parcel 8, a strip of woodland along Kenilworth Avenue near Crescent Road, to be part of a reconstituted greenbelt, and began efforts to purchase parcels 1 and 2 for the same purpose.

In 1984 Greenbelt residents experienced something new, which reflected the changing image of their town: the desire of people to join the city. The Hunting Ridge Condominium Association sought annexation to Greenbelt "to enhance the prospects for high-quality, low-density development in the area around Hunting Ridge."[79] The city annexed this area, south of the city limits next to the Beltway, along with Greenway Plaza Shopping Center and the Maryland Trade Center. The desire for this action by Hunting Ridge residents showed their belief in the statement by the city manager, James Giese: "By the '80s, the city had a great deal of influence over planning and zoning, political influence."[80] This marked a dramatic turnaround from the situation twenty years earlier.

The city government continued to emphasize the importance of both recreation and green space. The recreation department was one of four finalists in a 1983 competition for excellence in park and recreational management. In May 1985 a city bond issue passed by a four-to-one margin, providing money for the purchase of land to be maintained as green space as well as for construction of an indoor swimming pool and city building improvements. In June 1987 the city council wished to use money designated for the acquisition of parkland for building projects instead, as construction costs kept exceeding projections. An editorial against this appeared in the *News Review,* along with many letters to the editor insisting that the money be used for its original purpose. In July the city council promised that the full $2 million originally allotted would be used for acquisition of parkland.

City residents remained aware of the importance of the Beltsville Agricultural Research Station, to the north of town, as part of their greenbelt and monitored any attempts at change. Periodically, federal agencies made efforts to sell off part of the land. In 1987 an amendment to an appropriations bill, introduced by the U.S. congressman from Prince George's County, Steny Hoyer, prohibited the sale of Agricultural Research Station land as surplus.

Greenbelters have always been eager to tell others that Greenbelt is great. In fact, they are frequently spotted in warm weather wearing their "Greenbelt is Great" T-shirts. As part of a continuing effort to inform others of the truth of their slogan, in 1985 a Greenbelt team entered the All-American Cities competition sponsored by the National Municipal League and *USA Today.* The judges selected Greenbelt as one of twenty cities invited to give a presentation in Cincinnati. The team had ten minutes in front of the jury to describe "significant community-wide achievements." The Greenbelters chose A Tradition of Citizen Action as their theme. Naturally, a committee formed to prepare the presentation. The committee chairman hoped "to prepare a first-class presentation that truly depicts the tradition of citizen involvement in Greenbelt."[81] Even though Greenbelt did not win, the team returned from Cincinnati euphoric over the way their presentation had gone and happy to have represented Greenbelt.

Throughout the decades of development, Greenbelters not only demonstrated their pride in their community but also united to maintain their community as something to be proud of. They fought unceasingly against development that did not meet their criteria for what belonged in a planned, cooperative community. They also added community features they desired, such as a centrally located library, and parks and recreation areas. Working together on development issues helped forge community bonds that kept residents cohesive even as many forces in the period of the 1960s through the 1980s conspired to divide them.

# 6

# Overcoming Difficulties
# in Cooperation

THROUGHOUT THE POSTWAR PERIOD, while Greenbelters
fought developers in order to maintain their vision of a planned community,
their attention was continually distracted by other issues. Some of these
issues reflected turmoil in the larger society, such as the Vietnam War, while
others, such as the co-op Greenbelt Consumer Services, Incorporated (GCS),
remained peculiar to Greenbelt. These disputes often divided the townspeople into opposing camps, creating battle zones in a town in which people
took such matters very seriously. Indeed, if residents did not care so very
much, these issues would not have been so divisive. The major disruptions
involved the GCS, Greenbelt Homes, Incorporated, the Vietnam War, fair
housing, and a major dispute between residents of a new section of town,
Greenbriar, with residents of the central portion of town.

Over the years, the GCS greatly changed its presence, vacating its North
End store, the service station, the garage, the barbershop; the beauty, tobacco,
and valet shops; and the drugstore and variety store, while opening a new
"general store" in its expanded supermarket building behind the shopping
area. At the same time the GCS continued to expand outside the city, opening
a general store in Takoma Park, where a GCS grocery already operated.

In July 1956 most of GCS's 5,000 members approved two changes in its charter that proved to be significant. The new rules raised the limit on the amount of capital stock that could be issued from $1 million to $50 million and removed the $1 thousand limitation on the amount of co-op stock an individual member could own. In July 1957 the GCS opened its fifth major shopping center, a general store in Rockville. The following September, the president of Rochdale Cooperative, Incorporated, and the president of the GCS, Walter Bierwagen, announced the combining of the two operations. Rochdale continued to operate its stores in the District of Columbia and Virginia, while the GCS operated the Maryland stores. The combined cooperative now comprised eight shopping centers with more than 15,000 members. In 1959 the GCS moved its offices to Beltsville; only one board member lived in Greenbelt.[1] The owner of Greenbelt's original shopping center, Alfred Gilbert, who had purchased it from the government in 1954, rented his empty stores to new occupants. In 1958 the drugstore, variety store, service station, theater, and bowling alley came under new management. The central shopping area, once comprising exclusively co-op stores, filled up with private businesses.

Opinion in town became more divided as to the wisdom of the course pursued by the GCS. As always with controversies, large or small, people expressed their views in letters to the editor of the *News Review*. Most letter writers were critical of the GCS in its new role as large corporation. These excerpts are typical:

> Last week I attended another meeting of GCS which turned out to be another farce. The will of 23 congressmen [representatives elected by co-op members] prevailed on all issues except those marked on the proxy ballots. I submit that GCS has become too large in its present organizational form to be truly responsive to the members; that it is now effectively run by a self-perpetuating board.

> No amount of money (in the form of dividends and refunds) can make up to your members for losing control of their store in exchange for the dubious privilege of electing powerless congress members at an advisory area meeting in order to hold an "annual meeting" with 50 members present and 2,500 proxies out of 8,000 members. This is a co-op?[2]

On April 19, 1956, shortly after Walter Volckhausen left town, the paper "reprinted as a public service" a lengthy statement that Volckhausen had made on February 10 to the GCS. "Lest I be misunderstood," Volckhausen began, "my objective is to produce not animosities, but introspection." He charged that "the board has effectively silenced the voice of the membership" and concluded that the GCS was "little more than another big business. That is not what most of us were aiming for when we started GCS. It can be different, if the board will make it so. I deeply hope it will." Volckhausen followed this with specific recommendations to change the structure of the organization to be more responsive to members. With the *News Review*'s re-printing of Volckhausen's statement, the staff made the following disclaimer: "This statement does not necessarily reflect the position of the *News Review* with regard to these issues, but we are sure all cooperators will find it interesting and thought-provoking. The GCS board was invited to comment on this statement concurrently, but refused to do so."

By May 3 the position of the *News Review* had changed, as revealed in an editorial entitled "An Unfettered Tongue:

> The area membership meeting of Greenbelt Consumer Services clearly pointed out again that control of the "cooperative" has lapsed from the membership to management. . . . As with all organized groups, permitting the person or persons in power to gain greater and greater control eventually leads to full control. This permission was granted mainly by membership lethargy and disinterest. If the members have lost their voice in the operation of GCS, it is their own fault.

Suggestions followed on ways for members to reclaim power.

In letters to the editor, members and defenders of GCS had their say. In the June 7 issue, a board member charged: "The editors of the *News Review* and other GCS members and shoppers know that the Board meetings are always open to the membership. This has been a long-honored policy and the agenda of every Board meeting provides for member's statements, communications, complaints, suggestions, etc." He went on to attack the paper: "Every unfounded criticism as well as legitimate criticism—and there is legitimate criticism—finds its way into print. Oft-time the two are so intermingled that it is difficult to discover the line between the lie and the truth. This is irresponsibility of a high order, to say the least, it is deliberate distortion, to call it by its true name." Thus the differences of opinion between the GCS and the paper that had begun in 1950 continued apace.

The GCS maintained its refusal to place advertising with the *News Review,* thus placing the paper in financial jeopardy. A reader addressed this problem in the March 1 issue:

> I am willing to accept the thesis of GCS that it is not profitable to advertise in the Review—not directly profitable, that is, in a way that is measurable from week to

week. It becomes then a question of whether they should advertise as a public service. Do they have any responsibility to Greenbelt beyond that of selling us our "money's worth?" We all, including the GCS management, surely agree that they have some. How much? Can they accept enough responsibility to see that the paper continues to exist, so that local problems can be aired and a body of enlightened citizen opinion continue to guide our community? I hope so, because if cooperatives, represented by GCS, cannot accept this much responsibility in a democracy they are lagging behind private business and are less ethical than these brothers in private enterprise.

An August 9 editorial addressed the relationship of the *News Review* with the GCS:

Another issue of the Greenbelt *News Review* has gone to press without containing any advertising from Greenbelt Consumer Services. The local business that does over a million dollars worth of business here is still showing its discontent at public criticism by refusing to support the only community newspaper.

Your editor would like to take this opportunity to express his deep appreciation to all the staff members and to all our volunteer reporters representing local civic groups for their fine work in keeping a community newspaper alive. It is an accomplishment of great and wonderful magnitude. It is part of the spirit that makes this city unique, along with its good planning, and community services and democratic participation.

We urge all our readers and supporters to continue the battle to keep a free newspaper alive in Greenbelt.

In November 1959 the GCS and the paper's staff finally solved their differences, and the GCS resumed its advertising, aiding the paper's financial situation considerably.

No matter how much GCS members disputed specific facts, they acknowledged that the GCS had become a big business and had grown above and away from town life. Most of those holding an opinion did not approve of the change. The paper thus reflected town sentiment accurately and was accordingly supported by the residents in its differences with the GCS.

The failure of Greenbelters to maintain control over their co-op reflected a lack of sufficient interest on their part. Walter Volckhausen, among others, attempted to generate more widespread involvement in the GCS but did not succeed. A few activists, those having much interest in, and spending much time on, GCS affairs, controlled its operations, along with the hired managers. Most Greenbelters deplored the situation but did not organize in time—when the GCS remained small enough—to change the direction of events. For co-op enthusiasts, and many lived in Greenbelt, the GCS re-

Advertisement placed by Greenbelt Homes, Inc., in the
*Washington Daily News*, April 21, 1961. Courtesy Tugwell
Room, Greenbelt Public Library

mained a success story. For Greenbelters who believed that a major tenet of a cooperative community was having its institutions under local control, the evolution of the GCS away from Greenbelt reflected a failure on their part.

The GCS continued to expand throughout the 1960s. In 1967 the co-op purchased nine food stores from Kroger Company, and by 1970 the group owned twenty-two supermarkets, ten service stations, five pharmacies, and seven SCAN stores.[3] Co-op headquarters had been moved to Savage, Maryland, in the 1970s, removing all associations with Greenbelt except for the local grocery, pharmacy, and gas station. The name, changed in 1979 to Greenbelt Cooperative, Incorporated, remained the only other tie to Greenbelt.

Early in 1984 Greenbelters received an unpleasant surprise when the board of directors and management of Greenbelt Cooperative announced its decision to close its supermarket and gas station divisions. Greenbelters knew that the co-op had suffered economic losses in its period of rapid expansion in the 1970s, but the announcement of the impending sale stunned members. They "spoke out sharply concerning the secrecy of the decision-making process, exclusion of the membership from knowledge of the seriousness of the chain's financial problems, and the board's failure to comply with the directive passed by 57% of the members of the GCI Congress to study other alternatives in order to maintain these vital services."[4]

Some confusion resulted when, in response to the announced closing, two different groups formed in Greenbelt, with two different plans for rectifying the situation. Jim Cassels organized the Committee to Preserve Greenbelt's Co-op, which started meeting weekly in late January. Subcommittees formed to study various aspects of the problem, with the goal of taking over the local supermarket, drugstore, and gas station by a Greenbelt group. Another group formed the Save Our Services Citizens Committee, which attempted to halt the GCI actions by legal means. In response to an appeal from this group, the Circuit Court of Prince George's County ordered the GCI to temporarily cease divestiture.

The Committee to Preserve Greenbelt's Co-op moved quickly, selecting a name, Greenbelt Consumer Cooperative, and making plans to finance the purchase. The committee hoped to raise 20 percent of the amount from co-op members and 80 percent from bank loans, probably from the National Consumer Cooperative Bank.[5] In March the SOS (Save Our Services) group collected signatures on petitions to forward to the GCI, while the new co-op had already raised $40,000 from members and had received a $240,000 bank loan. At the end of March the court ruled that the GCI could proceed with divestiture—which it speedily did, selling the Greenbelt operations to the new Greenbelt Consumer Cooperative.

Jim Cassels explained how his group was able to act so quickly: "Helping to organize a new co-op in Greenbelt is relatively simple. You've got a great history to build on and an educated public. They know what a co-op is, it's old hat to them, so its not hard to start another one."[6] Thus Greenbelt's grocery and drugstore co-op came full circle. They began in Greenbelt during the earliest days, expanded into a big business with little relation to Greenbelt, developed money troubles, were sold, and began anew in town, under local control once again.

Another of Greenbelt's key co-ops generated much local controversy as to its proper operation. Greenbelt Homes prospered, but by the late 1960s, it became apparent that renovation, especially of the original "defense homes," was needed. Determining exactly what should be done and how it should be financed caused much contention. On October 3, 1968, an advertisement in the *News Review* called members to a special meeting, titled Planning for the

Future. Topics on the agenda were capital improvements, a building program, an acquisition program, a financing method, and the integration of these several programs into a master plan. The advertisement concluded: "All interested members are urged to attend to make comments and offer suggestions."

Thus began a series of meetings filled with acrimonious debate between Greenbelt Homes management, board members, and co-op members. While the management and board attempted to focus on a long-term improvement plan, many members only saw looming increases in their monthly co-op charges. Debate frequently degenerated to the personal level, filling an entire page in the *News Review* with the letters of various factions. An editorial in the December 10, 1970, *News Review* tried to bring coherence to the controversy:

> Now there can be honest differences of opinion as to whether some of these expenditures were needed this year or could be deferred to another year, whether the work being proposed could better be left to individual responsibility, or whether the corporation was making most efficient use of present expenditures. But the defiant mood of the membership ruled out constructive give and take exchanges on these issues. Once again the members themselves were the losers.

At the end of the decade, nothing had been resolved.

In 1972 engineering consultants stated that $4,700 per unit would be necessary for renovation, at a total cost of $7.5 million. By May 1973 arguments heated up among the 1,600 Greenbelt Homes shareholders over what renovations should be done, when, and how. Reflecting the varying opinions, eleven people competed for the five vacancies on the board of directors. By November 1973 the rising cost of oil for the 1,600 oil-heated units created serious financial problems, underscoring the need for change.[7]

In November 1976 the board of directors voted to apply for federal money for the renovation in the form of community development grants. In June 1977 Greenbelt Homes obtained $150,000 from the U.S. Department of Housing and Urban Development (HUD) for consultants to devise the best long-range rehabilitation program. In September 1978 more than 500 Greenbelt Homes members attended a special meeting to vote for or against a large-scale rehabilitation project, with the vote approving renovation, 273 to 140. Arguments continued, but another vote to approve a $6.4 million HUD loan, with another $5.2 million for capital improvements, passed in September 1979. This loan program, eventually amounting to $17.5 million, was the largest such loan approved by the federal government. After four years of

A "We the People" float echoes the stone relief by Lenore
Thomas on the front of the school/community center.
Three generations of the Barcus family represent all of
Greenbelt's "pioneers" and their descendants in the Fiftieth
Anniversary Labor Day Parade, 1987. Photo by J. Henson,
courtesy of the *Greenbelt News Review*.

work, from 1980 to 1984, electric heat had replaced oil, all units had new
windows, upgraded wiring, and insulation, and frame houses had new shin-
gles and new siding.

Even though it caused massive arguments, the renovation undertaken by
co-op members during the late 1970s and early 1980s illustrates the continuing vitality of the co-op ideal. Thirty years after its beginning, a new generation of members willingly undertook a major project to ensure the continued
long-term existence of Greenbelt Homes.

In February 1978 Greenbelt Homes members celebrated as they paid off
the original 1952 mortgage. One Greenbelt Homes veteran, looking over the
past, remarked, "There was a very real sense of accomplishment. There
weren't more than a dozen people outside the co-ops who thought we'd
succeed."[8] A celebration in honor of the retirement of the mortgage took
place in the auditorium of Eleanor Roosevelt High School. U.S. Representa-

tive Gladys Spellman, a favorite of Greenbelters from her days on the Prince
George's County Council, spoke eloquently:

> I can't think of a better place in which to hold this ceremony than in a building
> named for Eleanor Roosevelt. . . . She had real beauty, inner beauty. Beauty of the
> kind that is meaningful. She had a great fondness for Greenbelt and the concept
> which the original town represented. I don't find that surprising. You're her kind of
> people, and Greenbelt would be her kind of place. Better than most people of her
> background and generation, she knew the meaning of the word *community*. She
> would, I have no doubt, rejoiced with you that what began as a furrow in the brow
> of that eminent brain truster and New Dealer Rex Tugwell is ending up as a
> cooperative housing corporation owned and operated by the people themselves.
>
> Playwright Henrik Ibsen might have had Greenbelt Homes in mind when he
> wrote, "A community is like a ship: everyone ought to be prepared to take the
> helm." Indeed, back in the turbulent '60's, I understand, it seemed at times that
> everyone was determined to take the helm—sometimes all at the same time.
>
> But today is not the time to dwell on past battles; it is the time to remember that
> out of divisions and differences is born consensus, without which democracy at
> every level would surely falter. . . . To falter, but keep on going; to differ, but to come
> together again and to ultimately succeed in a common, shared enterprise—that's
> the stuff of which triumphs are made.[9]

Not all controversies were caused by the operation of local co-ops; some
difficulties reflected events in the larger society. Foremost among these was
the war in Vietnam. This issue caused particular problems due to the continuation of two differing, influential groups: liberals who tended toward
pacifism and the American Legion. By early 1966 the Greenbelt Committee
for Peace in Vietnam (CPVN) began to assert itself in town life. In many
communities differences of opinion could remain unfocused or unexpressed,
but Greenbelters pride themselves on the free expression of ideas. Thus
trouble erupted in that foremost of Greenbelt traditions, the Labor Day
festival.

Community groups, except those in partisan politics, had always been
welcome to sponsor booths at the festival. However, the local chapters of the
American Legion and the Disabled American Veterans (DAV) announced
that they would pull out of the festival if the CPVN participated. The chairman of the Booth Committee, after being told by several other groups that
they would also withdraw if the CPVN participated, made the decision to bar
the CPVN from the festival. She stated in the August 18 *News Review:* "In
defense of myself, I will state that I made the decision in favor of a majority.
My personal opinions concerning all parties involved will not be voiced." A

statement in protest of this action by the CPVN chairman appeared in the same issue.

Greenbelters fought the battle, as usual, through letters to the editor of the *News Review*. The August 25 issue contained letters with the following comments:

> If the presence of the peace group's booth offends individual citizens of Greenbelt, they can and should boycott the booth. If the presence of this booth offends the American Legion and the DAV, they are, as they have pointed out, perfectly free to withdraw. It is difficult for us to understand, however, how these two veterans' organizations, which have long and vociferously advocated the "American way of life," could allow the Committee to so blatantly contradict that very concept, as manifested in the healthy American tradition of airing honest differences of opinion in a peaceful manner.

> Two years ago when I brought my family to Greenbelt I had thought I had found the ideal place to live in the Washington area, primarily because of the type of people who live here—people who care about their community, people with ideas. My beautiful dream was shattered not by the rape of Greenbelt but by the action taken against the Greenbelt Committee for Peace in Viet Nam. . . . The Festival Committee, the *News Review* and the citizenry should publicly censure those organizations that threatened to withdraw from the Labor Day activities unless CPVN would be deprived of its booth. It is far too fundamental for discussion that democracy cannot work unless we afford those who disagree with us the right to dissent.

Instead of appearing at the festival, the CPVN held a peace rally at Center School.

On April 10, 1967, the director of the National Selective Service system addressed the Greenbelt Lions Club at the American Legion hall. A group carrying placards stating their opposition both to the draft and to the war in Vietnam picketed outside the hall, prompting concerned legionnaires to call the police. However, the Lions Club president invited the picketers to join the meeting. They did, listened, asked questions, and "the evening ended in an orgy of hand-shaking" according to the April 13 *News Review*.

As the time drew near for the 1967 Labor Day festival, the American Legion and the DAV refused to participate because the CPVN planned, once again, to have a booth. The August 17 *News Review* commented in an editorial: "Approximately 20 local organizations generally sponsor booths in the festival, and it does not seem reasonable that each of them should have to meet with the approval of the Legion's Americanism Committee. Is this really Americanism?" However, the CPVN voluntarily withdrew from the festival, "under pressure from many of the remaining participants, and in hopes of

Labor Day parade, Greenbelt, 1975. Photo by Nick Pergola,
courtesy *Greenbelt News Review*

preventing severe restrictions against all participants," as committee members explained in a letter to the editor. They again held a peace festival on
Center School grounds, which attracted police attention. Greenbelt police
appeared twice to take pictures of committee members, charged one member
with "soliciting without a license," and asked one member to come to the
municipal building, where he faced five city councilmen, the police chief, and
the director of public works. They informed him that the CPVN exhibit was
on city property and requested that it be moved several feet, to county
property.[10]

   As protests against the war spread, with many in the country agreeing that
the war was wrong, the issue ceased to be such a divisive one in Greenbelt.
At a special hearing on August 17, 1970, to discuss the McGovern-Hatfield
amendment to end the war, the city council decided not to take a position on
national matters, and the issue did not come to a vote. However, from coverage in the *News Review,* which usually reflected Greenbelters opinions accurately, it appeared that the CPVN now represented the feelings of many
Greenbelters.

Another important issue of the 1960s came to the fore during the same period as the Vietnam War, that of fair housing, but it did not openly divide the community as did the war. Liberal activists in Greenbelt, probably suffering from guilty consciences for living in an all-white enclave, brought the issue to public attention. Most Greenbelt residents regarded themselves as very tolerant of others and attempted to behave in a way that met with their self-expectations, as they lived out their ethic of cooperation.[11] Yet Greenbelt remained the all-white suburb constructed in 1937 when segregation was an unquestioned fact of life. The racial turmoil of the 1960s forced residents to reconcile this history of segregation with their self-image of openness and liberalism. Even before the civil rights era, Greenbelters tried to overcome the all-white reality of community life. African Americans who worked in town, for example, could shop at local stores and eat at the drugstore. They could not eat in a "white" restaurant anywhere else in Maryland or the District of Columbia.

Even though Greenbelt Homes never made a stipulation regarding race, by 1963 a small group of housing activists decided that positive steps should be taken to attract African Americans to Greenbelt. The group started the way that most groups did in Greenbelt, by taking out an advertisement in the *News Review* to announce its formation and inviting citizens to come to an organizational meeting. On October 17, 1963, more than sixty people, interested in seeing Greenbelt offer equal housing opportunity, formed a group and began work. In response to pressure from the housing group, the Greenbelt Homes board of directors explicitly reaffirmed the existing membership policy, which made no stipulations regarding race. An active member recalled his efforts:

> We never had to do anything except drop a hint. There was a couple, for instance, that applied for housing in the co-op. One of them worked for the FBI and the other worked for the Army, I think. We went to the Board of Directors who had to vote on the people coming in. We said, "You wouldn't dare vote against them, would you?" They didn't, of course. Well, they moved into a house on Southway and one of the neighbors there said, "By God, I won't live next to any niggers here. I'll move out first." But who helped this couple move in? This very same man, and I think he is still there. The couple has moved on to bigger and better things, I guess, but that was the closest we ever came to a confrontation. Nothing happened.[12]

Members of Greenbelt Citizens for Fair Housing attempted to actively recruit blacks into Greenbelt Homes housing by advertising widely and speaking to church congregations, particularly in the District of Columbia. Their efforts met with little success due to two problems: the image of Greenbelt as an all-white community left over from its federal government days, and the fact that blacks who escaped from the inner city often wanted to live in single-family homes, not the barrackslike row houses of Greenbelt Homes.[13] Thus the availability of new areas of single-family homes attracted more blacks to town than the outreach of Greenbelt Homes.[14] Besides working with Greenbelt Homes, the fair housing group sought legal changes that fell within the domain of the city council. In September 1967, for example, the council voted to ask the owners and managers of Greenbelt apartments to consider voluntary adoption of nonsegregated rental practices and policies. However, this action became unnecessary when the Prince George's County commissioners passed a fair housing ordinance that applied to the sale or rental of all housing in the county.

In actuality, the attitudes of Greenbelters probably encouraged integration more effectively than the specific efforts of fair housing supporters. According to a 1965 study of Greenbelt's fair housing movement, Greenbelters "illustrated that it was unfashionable to give voice to racial prejudice. Resident managers in particular were uncomfortable when put in positions where they had to acknowledge discriminatory policies. . . . This avoidance of being tagged as 'prejudiced' could have made the organization of resistance difficult, insofar as organization would have required the open expression of this unacceptable attitude."[15] In "cooperative" Greenbelt, with its emphasis on acceptance of all people, open prejudice was not tolerated. The insistence on liberal Greenbelt mores made the work of Citizens for Fair Housing largely accepted by the community as a whole. Census data show that although no blacks resided in Greenbelt in 1960, 230 did in 1970. The number of "others," presumably Asian and Hispanic, increased from 17 to 146. These numbers still reflect a very small share of Greenbelt's 1970 population of 18,199, but the percentage of minority groups increased greatly after 1970 (see table A.2).

As with fair housing, court-ordered busing effected Greenbelters but was not opposed due to the Greenbelt creed of being open and accepting of others. In January 1973 U.S. District Court judge Frank Kaufman ordered the transfer of 32,000 students to achieve desegregation in the Prince George's County schools. North End Elementary students living in Boxwood Village were transferred to John Carroll Elementary School in Landover, as were Center School pupils from Charlestowne Village and University Square. One hundred and seventy-five students from Springhill Lake Elementary School now attended Oakcrest Elementary School in Landover. Greenbelt Junior High School students, except those from Springhill Lake, were bused to Mary Bethune Junior High School in Chapel Oaks. Greenbelt senior high school students remained at Parkdale High School.[16]

In response to the court-ordered busing, Greenbelt parents formed a group to ensure that the mandated changes went smoothly. The January 1973 issues of the *News Review* provided detailed information on the schools that Greenbelt children would be attending. Letters to the editor offered every conceivable viewpoint on the busing of children. Many parents objected to having their children transferred during the school year. Irate residents of Boxwood Village claimed they bore the brunt of the desegregation effort even though they lived in an integrated area (11 of Boxwood's 199 families were black).

While many parents remained unhappy that their children had to leave Greenbelt for their education, the town, relying on its tradition of openness and cooperation, made an effort to welcome entering students. Mayor Richard Pilski declared, "I have never known Greenbelters to turn their back on anyone. I am counting on the fairness of our citizens." Greenbelt schools arranged open houses for new students and their families. The principal of Center School remarked that it was "important to show the new students and parents that they are accepted and made welcome."[17] The shuffling of students proceeded smoothly, despite initial fears. However, any achievement toward desegregation was nullified by the fact that from 1970 to 1983 the percentage of whites in the population of Prince George's County dropped from 80 percent to 45 percent.[18]

A new type of conflict in town occurred when, for the first time in Greenbelt, one area pitted itself against the rest. This happened as a result of building the new high school in a physically isolated area. Greenbelt students from the original part of town who walked to school needed to go a considerable distance out of their way and across Greenbelt Road in order to cross the Baltimore-Washington Parkway. Many students began to take a more direct route, dashing across the parkway instead, dodging speeding cars and trucks. Fears for student safety mounted, leading to a suggestion for a walkway across the highway. The most accessible and convenient location for a walkway brought it near Greenbriar's phase 1 condominiums. The city secured federal funds for a pedestrian overpass, but Greenbriar residents failed to

Save the greenbelt booth, Labor Day festival, Greenbelt, 1991. Courtesy *Greenbelt News Review*

approve a right-of-way. They objected to the presence of students walking through their neighborhood and feared that the walkway would bring in "undesirable elements." One resident said during a meeting, "We didn't move into Greenbelt. We moved into Greenbriar." Another resident, who objected to this line of thought, wrote to the March 23, 1978, *News Review:* "I thought this overpass would bridge more than a highway and make us 'closer' Greenbelters. But evidently there are Greenbriar residents who relish the physical barrier of the Parkway as preserving a sort of splendid isolation."

At meetings to discuss the issue, Greenbriar inhabitants claimed their area had the highest crime rate in the city and that the overpass would exacerbate the problem. However, city statistics for 1978 show that Greenbriar and adjacent Glen Oaks and Windsor Green had by far the lowest crime rate.[19] Many of those in Greenbriar favored a route further from their neighborhoods, which would have made the walk from central Greenbelt to the high school much longer. The president of Greenbriar's phase 3 condominiums wrote to the April 17 *News Review:* "It is our hope that the city council will reconsider its position and drop all plans to install the overpass in the Phase I location or any location within the property limits of the Greenbriar complex." Not all

Major renovation of original Greenbelt row houses,
Greenbelt Homes, Inc., early 1980s. Photo by Sandra L.
Henson, courtesy *Greenbelt News Review*

residents agreed with this view; the next week, in the April 26 *News Review*,
one responded:

> Being an owner of Phase III and a building representative, I am quite embarrassed
> by this and want to apologize in behalf of other Phase III owners who are in favor of
> this most needed overpass.
>
> I agree with the people of Old Greenbelt who would be affected by the overpass
> cutting a path by their homes, but yet aren't selfish enough to be concerned about
> crime and vandalism to not want the overpass. They are more importantly con-
> cerned with fellow Greenbelters' human lives.
>
> Face it, no one would have to be concerned with illegal crossing on the Parkway
> if this overpass were constructed. I want to feel a part of the Greenbelt community
> but unfortunately can't until this major link is completed.
>
> Come on, Greenbriar, stop being so selfish and paranoid, and start thinking
> rationally again! What will it take to convince you? Probably two dead children
> lying in a pool of blood on the parkway?

It is interesting that in the middle of the overpass dispute, being carried on
in traditional Greenbelt style with discussion at city council meetings and
letters to the editor, the residents of the Greenbriar Association gathered

enough proxy votes to outweigh the developer's votes at their annual meeting. The residents achieved this by door-to-door canvassing and letters to absentee owners. They took control of the Greenbriar Recreational Association, which handled utilities, recreational facilities, the clubhouse, and common areas for all of Greenbriar. Whether they realized it or not, their actions were typical of old Greenbelt, with which many did not wish to associate.

In May 1978 the city council voted to construct the overpass in the Gardenway location in Greenbriar, which was the shortest route from the Center to the school. When the city began condemnation proceedings on the necessary land in Greenbriar, through the median strip of a parking lot, Greenbriar residents took the issue to court. A jury awarded Greenbriar $29,000 for the land, but Greenbriar residents appealed the decision. The court of appeals overturned the lower court decision, stating that the city should have considered individual demands for damages from residents whose apartments were nearest the walkway.

A referendum of all city residents showed that 88 percent of the voters favored the Gardenway site for the overpass, and the majority prevailed after the city awarded a large sum to Greenbriar for the land. This controversy with Greenbriar regarding the overpass exhibited the first instance in which residents of one area gave their interests precedence over the general welfare. The fact that the Greenbriar area was the first "luxury development" to be built in town is perhaps pertinent. After much disagreement and delay, by June 1983 students walking to Eleanor Roosevelt High School from old Greenbelt could use the pedestrian overpass instead of crossing the parkway.[20]

Throughout the 1960s, as Greenbelters coped with such insistent divisive factors, they attempted to maintain town traditions and continue with life as usual. The continuation of such traditions almost always proceeded smoothly but occasionally led to controversy. In 1964 city officials made the decision to cancel the Fourth of July festivities as an "economy measure." It also appeared that the Labor Day festival might not be held. As city finances actually remained stable, it appeared that the real reason was that city officials were overwhelmed by the legal challenges involving city land and could not expend the effort necessary to hold the usual celebrations. The staff of the *News Review,* usually quick to notice and comment on such deficiencies, had their hands full covering the continually changing development situation and made no mention of the festival. However, a *News Review* reader took over for them in the July 9 issue:

I have been a resident of Greenbelt for over twenty-five years and I am shocked at the way the two big holidays have been eliminated from Greenbelt's activities.

For many years, the Fourth of July Parade and the fireworks have been a real pleasure, not only to Greenbelters but to other people in the area. Most people cannot believe this has been eliminated.

As for the three day celebration [Labor Day] I think it is disgraceful that we're not having this successful affair. It is known that some former Greenbelters return to Greenbelt on their vacation just for the celebration. Everyone looks forward to it, the young children, the teen-agers, and older people.

I have raised four children, who now have their own families and they never miss coming to Greenbelt for the holidays.

The three day celebration is also a big boost to our merchants, as people come from the surrounding areas to the activities and while here, make many purchases at our stores.

I would like to close on a very disappointing note and say that I hope Greenbelters will come forward. I know it's too late to do anything about the Fourth of July, but please think of all the pleasures you will be missing if we do not have the Labor Day celebration.

After this letter appeared, Greenbelters did "come forward," and the 1964 festival occurred as usual. By 1968, when it appeared that Greenbelters could successfully hold out against developers, the festival attracted more attention than in recent years. Parade marshals David Eisenhower and his fiancée, Julie Nixon, provided the highlight of the 1968 extravaganza, as a crowd, estimated at up to 15,000, lined the parade route. The young couple remarked favorably on the nonpartisan spirit of the crowd. The September 5 *News Review* exclaimed: "This was a Festival to remember! The often referred to 'spirit of Greenbelt' glowed in tangible form throughout the gala weekend—in the smiles of the good-humored crowds, in a pulsating sense of community. The harmony was relaxed and unforced. It drew together persons of all political persuasions, of diverse interests and enthusiasms."

In general, community traditions acted to unify Greenbelt, healing differences caused by disagreements over matters such as the proper management of Greenbelt Homes or political differences regarding the war in Vietnam. The persistence of these traditions enabled Greenbelt to maintain its identity, a topic examined next.

# IV

## A Town for the Ages

# 7

# The Persistence of the Greenbelt Idea

CITY TRADITIONS serve as the glue holding the town of Greenbelt together. These traditions, begun by city officials at the urging of the Farm Security Administration, unite the city by reminding one and all that they are Greenbelters.

During the 1970s the Greenbelt *News Review*'s readers successfully took into their hands one of the paper's previous responsibilities, that of remembering and encouraging town traditions. In fact, residents not only carried on with the traditional holidays and events but also created new ones. The spontaneous creation of new communal activities at this time could have emanated from a feeling that Greenbelt's unique identity was slipping away. With no causes to fight for as had been the case during the 1960s, and with the continuing encroachment of modern suburbia, town customs could serve to set Greenbelt apart as a special place.

A good example of such a custom was created by Leo Gerton, longtime manager of the High's Dairy store in the Center and an avid fisherman who sponsored the annual fish fry. This event began in 1964 on a small scale but by the 1970s had become a community event, with everyone invited to come to the lake park and to "bring a covered dish and condiments." The fish fry usually occurred in late summer or early fall.

The Festival of Lights began in 1972 as a modest Christmas crafts show and sale, along with caroling and a decorated tree. It eventually included concerts, decorating contests, and Santa's visit and lasted through the month of December.

In 1971 a resident came up with the idea of presenting an award to the Outstanding Citizen of the Year at the Labor Day festival. The award was to be "a way of recognizing what is special about this community—its high degree of voluntary service and cooperative action."[1] In typical Greenbelt fashion, a committee decided the procedures to be followed and presented the first award at the 1973 festival. Nominees, who have to be at least eighteen years of age, continue to be judged on "service to the entire community." The identity of the outstanding citizen is kept secret until the presentation, adding an air of suspense to the festival's opening ceremonies. The honoree then acts as grand marshal for the Labor Day parade.

Another city tradition began with the celebration of the first Greenbelt Day in 1975. In 1974 a committee formed to devise ways to celebrate the bicentennial in a fashion meaningful to Greenbelt. The committee decided to honor the issuing of the city charter on June 1, 1937, with an event to be held yearly on the first Sunday in June; the event was the Greenbelt Community Ball, held on May 31, 1975, at the National Guard Armory. The May 29 *News Review* helped to establish the new holiday in its usual fashion, with an editorial titled, "You Can't Afford to Miss It." A front-page headline announced "Greenbelt Day Activities Will Last All Weekend," while the accompanying article described the new holiday as "a time for all Greenbelters to commemorate Greenbelt's unique past, to contemplate the city's future, and most especially to celebrate what we have here today, including numerous hard-working citizens and good friends who have helped make Greenbelt a nice place to live." The paper ran excerpts of articles from the Charter Day edition of 1938, as well as a series titled "40 Years Ago."

The creation of new Greenbelt traditions in the first half of the 1970s filled parts of the calendar that heretofore had no special Greenbelt connection. Events in June and December, added to the major festival held on Labor Day weekend, helped keep the spirit of Greenbelt alive throughout the year. Why did this increased emphasis on the uniqueness of Greenbelt occur at this time? After the settlement of the Supreme Court suit and the reformation of county government in 1970, the massive efforts to preserve the identity of the community were no longer necessary. However, these intense periods of effort not only defined the community but also helped maintain it as a viable entity. Without the calls to action, the special qualities that define Greenbelt could gradually fade, until it resembled any other community. While no one articulated such feelings, they must have existed, leading to conscious efforts to keep Greenbelt a unique place by the creation of year-round reminders.

Greenbelt's premier event, the Labor Day festival, continued in a grand fashion throughout the 1970s. The final draft of the schedule of events for the 1971 festival contained eight pages of events over a four-day period.[2] In 1979, on the twenty-fifth anniversary of the festival, an editorial in the August 30 *News Review* commented on the occasion and addressed new Greenbelters: "The newer residents of our city, especially those from Greenbriar, Windsor Green, and Glen Oaks, should not miss out on the chance to be a part of their new home town. We strongly urge them to come out and join the rest of us for our four-day party." In election years, politicians take advantage of the crowds to put in an appearance. Steny Hoyer, the U.S. representative from Maryland's Fifth District, maintains close ties to Greenbelt and usually appears at the festival.

Greenbelt's original ideals persisted in ways other than the creation and maintenance of town traditions. Co-ops remained important, not only in business, such as the grocery store, and housing, such as Greenbelt Homes, Incorporated, but also in meeting other town needs. New co-ops continued to be created when residents felt it beneficial. Advocates of a cultural center decided to utilize the vacant Utopia Theater and formed the Greenbelt Utopia Theater and Arts Cooperative. In November 1979 they launched a door-to-door fund-raising drive to pay for the theater's lease. By January 1980 the group raised enough money to sign the lease, with the Greenbelt Players presenting their first efforts in March. Some early productions were a Broadway musical revue and the plays *The Owl and the Pussy Cat* and *The Glass Menagerie.* Eventually, offerings expanded to include art classes for both adults and children. A grant from the Maryland State Arts Council provided firm financial footing.

Another co-op formed to meet the housing needs of Greenbelt's aging residents. Interested volunteers and city workers created Green Ridge House, financed by a state loan and subsidized by federal funds. Dedicated on June 3, 1979, the building on Ridge Road, with its one hundred apartments, dining room, workrooms, lounges, and library, is easily accessible by foot to the Center. At the dedication ceremony, State Senator Edward Conroy linked Greenbelt to the New Deal programs of a "disabled" president and spoke of the "appropriateness of establishing here the first home for the elderly and the disabled to be constructed under the Community Development Administration of the Maryland Department of Economic and Community Development."[3] The city took over the administration of Green Ridge House once it began operation.

Another volunteer effort to meet a community need involved the creation of Greenbelt Cares Counseling Center, which began as a one-night-a-week walk-in counseling program housed at the Greenbelt Community Church. State funds allowed the opening of the Youth Service Bureau, while the city applied for and received federal funds so that family counseling could continue on a larger scale. From its extremely modest beginning in 1972, the program grew until in 1978 it had five paid staff members, more than twenty counseling volunteers, and a budget of $57,000. The program is housed in the municipal building and is paid for by a combination of state and city funds. The city, on its own, acted to meet a continuing demand for recreation programs, building Springhill Lake Recreation Center, which opened in September 1975.

A new "cause" in town originated in the Earth Day celebration of 1970. In 1971 articles and letters to the editor of the *News Review* regarding recycling, pollution, and other environmental issues appeared regularly. The city began and maintained programs for recycling newspapers. Greenbelt residents had always considered trees sacred, but concern increased, and discussions on where to place new sidewalks to cut down the least number of trees sometimes became quite heated. In 1978 at a public hearing on the plan for the Greenbelt Metro station, residents objected strongly to the current design geared only to the automobile. They requested changes to allow access to the station for pedestrians and bicyclists. The Greenbelt Clean-up, Fix-up, Paint-up Day, begun in May 1976, became a regular occurrence.

Greenbelters revealed their organizational acumen on the state level in 1985 during a financial crisis. The origin of the problem occurred in 1973 when the Maryland Savings–Share Insurance Corporation began insuring Twin Pines Savings and Loan, a Greenbelt co-op. Greenbelters regarded their inclusion in this program as a historic event, because for the first time loans made to purchase homes in a co-op were "recognized by a major institution of the financial community as a legitimate investment."[4] However, the severe financial difficulties of several Maryland savings and loan institutions in 1985 led to the collapse of the state savings and loan insurance system. In late summer Maryland banking officials froze assets throughout the state, including Greenbelt's own Community Savings and Loan (the new name for Twin Pines). In response, the next issue of the *News Review* contained a letter inviting concerned residents to a meeting. Within a month Greenbelters had formed a statewide organization. For the next five months they organized meetings and protest marches until the state legislature dealt with the problem. Mellon Bank of Maryland acquired the assets of Community Savings and Loan, opening their office in the Center in May 1986.[5]

The manner in which Greenbelters responded to this crisis is typical and best described by a resident who was involved:

> When I moved to town in 1985 I put my money in Community Savings and Loan, which collapsed. I responded to a two-line notice in the *News Review.* The meeting was attended by 80–90 furious Greenbelters. But they weren't just angry. They knew what to do. Within an hour and a half they had formed an organization: the Maryland Savings and Loan Depositors Association. A month later it was statewide. The next meeting I went to, at Eleanor Roosevelt High School, had 700–800 people from all over the state. A month later Harry Hughes made the mistake of visiting the new Greenbelt Hilton to make a speech, and he had 300 people picketing, circling the Hilton. All of this was people who knew exactly what to do. People who were hell-raisers from 30–40 years ago and they knew how to run a community organization, make it go and give the politicians hell.
>
> They spread out throughout the state. It was all centered in Greenbelt. This was all something that would not have happened in any other suburb of Washington. Maybe Takoma Park. Greenbelt knows how to organize. You could just smell all the old battles all these grizzled veterans had been through, with any number of levels of government. . . .
>
> This community has echoes of those New Dealers. I can see the ghosts of a lot of those folks smiling.[6]

This quotation reveals how much Greenbelt's history remained in the present. Residents retained more than just the traditions of the city. Knowledge of how to start a co-op, how to mobilize large numbers of people quickly, and the most effective ways to lobby politicians had been passed down from early activists to those working for their town in the mid-1980s. The town youth obviously learned from observing their elders. Throughout the 1970s, police periodically chased groups of teenagers away from the Center, accusing them of loitering. These young people appeared at city council meetings to protest police actions, saying that adults "loiter" at the Center all the time. One adult observer wrote to the October 23, 1975, *News Review* that the fact that the youth "chose to come before their elected officials rather than take the vigilante route is the healthiest sign of government in action you could find. . . . A city council meeting is Government of the People, by the People, and for the People. That premise is alive and working in Greenbelt. Be grateful."

The original ideal of cooperation in Greenbelt meant continued cooperation among Greenbelt's religious groups. Interfaith Thanksgiving Day services were held regularly. Emulating other sectors of Greenbelt life, community religious leaders formed the Greenbelt Clergy Association to coordinate activities and seek ways to improve town life. In November 1976 the association launched an effort to reach apartment and condominium dwellers physically separated from the core of the city. It felt that the best way to involve these people in the life of the city would be to improve the circulation of the *News Review* to outlying areas. The association worked with city officials and local organizations to develop a welcome kit for new town residents.[7] Throughout the 1980s the tradition of interreligious cooperation in Greenbelt continued, receiving a boost with the establishment of an interfaith walk in April 1987 to mark several religious holidays. As the *Washington Post* reported:

> An unusual gathering of Protestant clerics, a Jewish rabbi, Catholic laypeople and Baha'i believers undertook a trip through the planned community of Greenbelt yesterday to celebrate the confluence of three distinct religious holidays being observed this week. Saying they were reflecting "on the brokenness of our world," the Greenbelt Clergy Association and a group of 40 made 14 stops for meditation or prayer. The trip included stops at a formerly troubled savings and loan and a children's playground that residents say is now used by drug pushers after dark. . . . The clergy association decided this year to expand observance of the Christian holy day of Good Friday to include the Jewish Passover season and Ridvan, the 12-day celebration of the founding of the Baha'i faith.[8]

The *Post* writer continued, "Cooperative ventures, religious or secular, come easily to Greenbelt, the 50-year-old planned community established under Franklin D. Roosevelt's New Deal. Envisioned as a garden community built by and for the victims of the Depression, it has evolved today as a middle-class bedroom suburb, but one with a strong sense of community and social responsibility." In other words, Greenbelters persisted in keeping the original goals of the city in view and in retaining an active interest in city affairs. Rhea Cohen, for example, the head of Save Our Community Committee, ran for and won a place on the city council, following the common pattern of first spearheading local concern on a particular issue and then seeking political office.

In December 1977 the city learned that a proposal by the Prince George's

County Board of Supervisors of Elections would rearrange and split Green-belt precincts for general elections. City council members objected strongly, as they felt the changes "would destroy the integrity of Greenbelt's bound-aries and lead to a lack of political cohesiveness within the community."[9] The board of supervisors stated that Greenbelt was not consulted because city elections were not involved. However, city officials knew that being assigned to two different polling places would be confusing to residents. They also understood that Greenbelt's identity depended, in part, on the maintenance of boundaries like voting districts.

In view of local unrest, election supervisors conducted a hearing in Green-belt to receive local opinion. According to the December 15 *News Review,* "the hearing before the Board of Supervisors of Elections of Prince George's County the evening of December 5 was the most nearly unanimous meeting of any kind that this reporter has ever attended in Greenbelt." City council-man Charles Schwan summed up local feeling: "The sense of community is greater than in most municipalities of similar size and this despite the fact that the city is divided into nine segments by major highways." He asked that a new polling place for Greenbelt East be for Greenbelt residents only, to continue city cohesiveness. In April election officials announced that they would leave Greenbelt precincts alone but that they would add one precinct in Greenbelt East, which would include some non-Greenbelt residents. The determination of Greenbelters to maintain their electoral boundaries illus-trates their vigilance regarding threats to the integrity of their town. The fact that city residents controlled their own affairs continued to contribute to the city's identity, and residents wished to have no confusion regarding who was and was not a Greenbelter.

In 1981 Greenbelt residents again disputed voting districts, this time the redistricting of county council seats. Greenbelt had asked to be in a different district from Bowie, which had three times the population of Greenbelt. The new plan, a mirror image of the historic gerrymander in Massachusetts, placed Greenbelt in a district with Bowie and Upper Marlboro. Greenbelters, more than eighty strong, appeared at the county council meeting in Upper Marlboro to voice their objections. U.S. Representative Steny Hoyer sent a letter to the council chairman on behalf of Greenbelt, and Greenbelters gathered more than a thousand signatures on petitions, all to no avail.

In addition to maintaining the town boundary and actively running their city, Greenbelters increasingly favored the marking of anniversaries as a way to focus on and remember town origins. Residents representing almost every organization in town attended a city council meeting in March 1962 to plan a twenty-fifth anniversary celebration. A steering committee formed, with a subcommittee for each proposed event. Plans blossomed and came to fruition in October. The American Legion Post held a luncheon for the "first families" of Greenbelt, and Greenbelters danced at a gala anniversary ball. The *News Review* printed three special editions detailing the history of the town and its numerous organizations; these were later published as a booklet. The November 15, 1962, *News Review* summarized the activities this way:

> As always, the watchword was cooperation—the voluntary associations of public-spirited citizens to perform a service and to fill a need. Under the guidance of chairman Harry Zubkoff the Twenty-fifth Anniversary Committee turned in an exceptionally able piece of work. Everything went smoothly, everything was fun— the special luncheon for the first families of Greenbelt, the Space Agency Open House, the anniversary ball. In addition, an able job of public relations saw to it that all the metropolitan papers took note of our anniversary. Throughout the summer, the people we have met have been glowing proudly over the part they have played in Greenbelt's history.

The staff made sure that the *News Review* did its part in recalling the history of Greenbelt. A series of articles featured interviews with early town residents; the paper held a dinner for all those who had worked on the paper during its twenty-five-year history; a front-page article noted the death of Eleanor Roosevelt; and a series called "Twenty-five Years Ago" was created. The paper frequently recalled town history in articles on subjects as miscellaneous as the statue in the Center, the co-op nursery school, and the old burial grounds. A special November 30, 1967, supplement, produced for Greenbelt's thirtieth anniversary, provided opportunity for reflection:

> Thirty years ago our first issue appeared. Thirty years, one entire generation; and as one looks back at the yellowed pages of the early issues, the names one encounters indeed belong to a different generation of Greenbelters. Those now associated with the *News Review* might recognize only a few—perhaps none at all—of the pioneers who in 1937 started the "Greenbelt Cooperator." It is therefore the more remarkable, the more gratifying and wonderful, that our newspaper has not only existed these 30 years, but has also followed its founders' design and their original goals, as a creative and cohesive force in Greenbelt's community.
>
> A newspaper is a rather temporal thing. It is put together on short order—in our case, by amateur newspaper workers coming from a full day's work elsewhere; and though we always try to be as informative and comprehensive as we can, we harbor no illusions as to how much of this week's issue will be remembered a month hence. Sometimes it appears as if we live from one crisis to another—editors

Opening ceremonies, Greenbelt Museum, October 10, 1987.
Photo by Bill Cornett

leave to raise families; balance sheets dip into the red, at one time we even were threatened with eviction from our basement—and through them all, we are busy preparing the next issue, reporting the next meeting. Only rarely can we allow ourselves the luxury of stepping back and contemplating the large picture in which each effort is just a small contributing part.

Today is one of these rare occasions. As we look back today from the vantage point of 30 years, we cannot help admiring the foresight, in a community only six weeks old, of those who saw the need for a newspaper as a local rallying point. It is their enduring vision which is now giving strength to a new generation of writers and editors to carry on the task which they began.

Town history received attention in other ways also. The June 20, 1968, issue of the *News Review* noted on its first page:

The Greenbelt Historical Society cordially invites Greenbelt residents and any other interested parties to attend its first general meeting and to be active members in this newly formed organization. The City of Greenbelt, now in its thirty-first year, remains unique among American cities in its conception and original development. It is the intent of the Greenbelt Historical Society to record and preserve the history of this city while there are still many who can remember the early days of the city and can relate their experiences as first citizens in a model community.

On June 23, at the first meeting, about sixty people viewed photos of early Greenbelt and heard reminiscences from early inhabitants.

In 1970 the Prince George's County Historic Sites Identification Map added three Greenbelt locations: Indian Springs, Walker cemetery, and original

Greenbelt. Walker cemetery contains the graves of Isaac Walker and his son Nathan, who fought in the Revolutionary War. Original Greenbelt is called "a development potentially historic by virtue of its having been the first experience by the Federal Government in providing planned community housing for the low income level."[10] Indian Springs served as a gathering place for Indians from the time of the first white settlers; arrowheads can still be found there. The building of the Beltway destroyed public access to the site, which is reachable only by climbing through the storm sewer under the freeway. At the time of its addition to the Historic Sites Map, an article in the February 26, 1970, *News Review* commented on the irony of designating a historic site that could not be visited. The designation of historic sites in town and the formation of the Historical Society demonstrates that more than thirty years after the formation of Greenbelt both Greenbelters and others remained interested in preserving its history and meaning as a planned community.

Preservation efforts continued into the 1970s, as evidenced by the placing of a plaque at the Southway entrance to town during the 1975 Fourth of July celebrations; the plaque explained the historic importance of Greenbelt as a planned city. In 1979 Professor Dave Fogel, assistant dean of the School of Architecture at the University of Maryland, made arrangements to microfilm and preserve the original blueprints, which had been kept by Greenbelt Homes.

In 1980 Greenbelt was listed in the National Register of Historic Places. The historic district included the original housing, the Center, parklands, athletic fields, schools, churches, Indian Springs, several family cemeteries, and the remaining greenbelt to the city's north and east. The official declaration states: "The district is historically and architecturally significant for its association with community planning as one of the federal government's early experiments in planned towns. Greenbelt retains its basic integrity of design, materials, setting, feeling, and association. It has achieved exceptional significance dating from its construction in 1937 and 1941."[11]

During the 1970s a number of organizations celebrated anniversaries, continuing the focus on Greenbelt's history. In 1977 the Community Church celebrated its fortieth anniversary, and in 1980 Mishkan Torah its twenty-fifth. In 1980 Greenbelt Cooperative (formerly Greenbelt Consumer Services) and Greenbelt Nursery School reached their fortieth anniversaries. Also celebrated in 1980 were the twenty-fifth anniversary of the volunteer fire department and rescue squad and the tenth anniversary of the Greenbelt Library, which drew 250 people to its "party."

The greatest number of special celebrations planned during the 1970s occurred during the bicentennial. The city council established the Greenbelt American Revolution Bicentennial Committee to serve as a steering and coordinating committee. In November 1974 the *News Review* published a list of possible activities, many of which were accomplished. Several of the most difficult, such as the establishment of a museum, the histories of local organizations, and a conference for planners, occurred at the time of the fiftieth anniversary celebration instead.

The Bicentennial Ball, part of the second annual Greenbelt Day celebration, began the year's special activities. Events took place throughout the entire weekend, many of them similar to those of the Labor Day festival. Additional bicentennial events occurred on Fourth of July weekend as well as during the Labor Day festival.

A permanent addition to the documentation of Greenbelt history resulted from the bicentennial celebration with the creation of the "Greenbelt Trail Guide." Modified at the time of the fiftieth anniversary celebration, it is now available at the city hall and the library. The sixteen-page booklet provides a brief history of the city and a walking guide, with a map, to points of interest.

Soon after recovering from the bicentennial activities, residents began to organize for a 1982 celebration of both Greenbelt's forty-fifth anniversary and the centennial of Franklin Roosevelt's birth. In April 1982 the Greenbelt FDR Commemoration Committee met to decide how to celebrate, sending letters to city organizations inviting their participation. In May the committee considered various proposals to rename Greenbelt landmarks after New Deal figures. Suggestions included naming the lake park in Roosevelt's honor, changing Southway to Tugwell, and renaming the Center after Edward Filene, whose foundation provided money to initiate the co-ops. After much discussion over the following months, decisions were made to name the lake park Buddy Attick Park, after a beloved member of the first police force, while the Center officially became the Franklin Roosevelt Center.

The latter decision prompted a resident to protest in the usual form of a letter to the editor, saying that the Center looked too bedraggled to honor FDR properly. Another resident responded to this letter the following week, October 14, 1982:

The Center has remained the heart and hub of this one-time federally planned community and stands as a constant reminder that we are a city born of the 1930s, rich in history and tradition that was made possible by the vision and initiative displayed by Franklin Delano Roosevelt. Therefore, I applaud the city council's decision to approve the naming of the Greenbelt Center in honor of Mr. Roosevelt, for, among the alternative city facilities proposed for this purpose, there is no more representative site which so appropriately befits the occasion.

The actual celebration of FDR's centenary and Greenbelt's forty-fifth anniversary occurred during October 1–3, 1982, the highlight being a reception for "first families" and the reunion-dance. Sponsored by the Greenbelt Lions and Lion Belles, dancing to the sounds of Benny Goodman and Glenn Miller, old-timers mingled with current residents, exchanging stories. Attendees considered the weekend a great success, providing encouragement to those who hoped to have a much grander celebration for Greenbelt's fiftieth birthday party.

Even while plans for the fiftieth anniversary celebration moved forward, the community continued to emphasize its history. A celebration on October 11, 1984, marked the hundredth anniversary of Eleanor Roosevelt's birth. As the October 11 *News Review* proclaimed, "Centennial Festivities Mark Greenbelt's Special Feelings for Eleanor Roosevelt." In November 1984 Greenbelt Homes offered a three-evening series on the history of Greenbelt. The only school in the nation named for Eleanor Roosevelt continued to offer courses with emphasis on science and the arts. Special programs at the school celebrated its tenth anniversary in 1986. Both Caroline Ware, a colleague of Mrs. Roosevelt's who combined work as a cultural historian with an interest in international development, and Mrs. Roosevelt's granddaughter, Anna Eleanor Roosevelt Johnston, spoke at the final program.[12]

Throughout the 1970s and early 1980s, Greenbelters spent much time and energy on celebrations and anniversaries, then put this experience to good use planning events for the fiftieth anniversary celebration, beginning in 1983. The scope of their achievements followed from both their experience and their advance planning, allowing enough time for busy volunteers to cope with daily responsibilities and still be a part of Greenbelt's festivities. The manner in which these events were planned and accomplished demonstrates the manner in which Greenbelters continued to organize themselves, repeating patterns begun during the years of World War II. On July 11, 1983, the city council passed a resolution "to establish a committee to plan and organize the celebration of the fiftieth anniversary of the incorporation of the city of Greenbelt, Maryland."[13] The wording of the resolution that followed illustrates key features of Greenbelt life: the remembrance of its origin as a planned community, the participation of as many as possible to plan special activities, and the coordination of events by a committee, to be composed of five to twelve members, selected and appointed by the city council. The city manager and one member of the city council served as ex-officio nonvoting members. The committee itself selected a chairman and vice chairman. The Fiftieth Anniversary Committee was authorized to establish subcommittees responsible for specific tasks or programs and to select subcommittee members.

The committee elected Sandra Barnes as chairman and Al Herling as vice chairman at its organizational meeting on March 6, 1984. The group promptly sent letters to all known local organizations, asking for their participation and suggestions for the celebration. As its first event, the committee sponsored a booth at the 1984 Labor Day festival. Its Greenbelt Trivia Game drew interest, with winners receiving "Greenbelt Expert" buttons. The committee also sold Greenbelt flags, the proceeds going to support the committee's activities. By November seven subcommittees had been formed, their names revealing the events and actions that residents were interested enough in to actually carry out: Development of the Greenbelt Museum, Production of the Fiftieth Anniversary Book, Commissioning a Statue of Franklin Delano Roosevelt, 50th Anniversary Dance, National Conference on Planned Communities, House and Garden Tours, Fund-Raising, and Sales and Promotions. These subcommittees illustrate that remembering and recording Greenbelt's history was the primary focus of the celebration, with a secondary emphasis on having a good time. Additional subcommittees worked on the oral history project, tree planting, and the production of a slide show. The only subcommittee that did not accomplish its goal was the subcommittee to commission a statue of FDR, as another group, which was trying to establish a museum, persuaded the FDR subcommittee not to compete with it for scarce city funds.[14]

The Fiftieth Anniversary Committee met monthly to receive reports from its subcommittees and to coordinate tasks. In January 1985 the committee held a special meeting to receive ideas from local organizations and to encourage them to sponsor specific events or projects.[15] Some of the ideas for projects had been floating around for years, and the impetus of the celebration brought them to the fore. This was particularly true of the museum, which was first suggested in a letter to the editor from Dorothy Sucher, published in the July 13, 1972, *News Review*:

> Greenbelt was built during the Great Depression as an expression of the philosophy of the New Deal. What better place could be found in which to create a museum devoted to that significant period in our nation's history? . . . How one goes about creating a museum, I have no idea. But perhaps there are Greenbelters with experience in this field who would have suggestions. It seems like a tall order—and yet every museum that exists was started by somebody, sometime. Anybody interested?

The idea was discussed at the Greenbelt Historical Society December 1977 meeting. The society asked for donations of items appropriate for a museum, and in June 1978 a "first settler" donated original furniture that the city stored with the thought of someday having a museum in which to place it. In January 1985 the Museum Subcommittee began work to determine not only what kind of museum was wanted but also how to realize it. In April the subcommittee presented a packet on their goals and needs to the city council. Its statement of purpose was as follows:

> The objective of the Greenbelt Museum is to preserve, for the benefit of the public, Greenbelt's unique heritage as one of America's earliest planned towns, built during the Thirties as an expression of the philosophy of the New Deal and the Cooperative Movement, and as an experiment in community living. The Museum will celebrate the founding of the City, and display objects of historical and artistic interest related to that period, in a manner that will provide visitors with cultural and educational enrichment. Involvement of the community on an ongoing basis will be an integral part of the Museum.[16]

The subcommittee created a new group, Friends of the Greenbelt Museum, to raise the funds necessary to start the museum and then to see to its long-term needs. By March 1986 more than 170 people from all sectors of Greenbelt had joined the Friends of the Greenbelt Museum. However, problems surfaced; the only controversy surrounding the anniversary celebration was the one that erupted in July regarding the projected museum.

The Museum Subcommittee and Friends of the Greenbelt Museum asked the city to pay for a Greenbelt Homes unit to house the museum, with the Friends responsible for all further expenses. The issue was the subject of extensive and heated discussion at several meetings, and many letters, both for and against, appeared in the paper. Everyone agreed that the city should have a museum, but a number of residents did not feel the city should pay for it. However, the majority seemed to favor the proposal. In November the city council voted five to nothing to purchase the unit from Greenbelt Homes. This illustrates once again the close working relationship between the city government and volunteer groups. Once the city purchased the building, the Museum Subcommittee formed committees of its own: the Building, Collections, Exhibit/Tour, Fund-Raising, and Public Relations subcommittees.

Another idea ruminated on for a number of years that came to fruition at the fiftieth anniversary celebration was the collection of oral histories. The Greenbelt Historical Society discussed this project at its December 1977 meeting but did not proceed with the project at the time. Dorothy Lauber, executive secretary to the city manager, began collecting stories from original residents at the time of the forty-fifth anniversary. From 1981 to 1985, as part of their course work, students from the University of Maryland conducted interviews of early Greenbelters. As the fiftieth anniversary drew near, Leta Mach, editor of the Greenbelt Homes newsletter and already doing a series on Greenbelters, asked the city government to videotape her next interview. Those involved were so pleased with the results they did a series of videotaped interviews of early residents as part of the celebration. This oral history project then became a subcommittee of the Fiftieth Anniversary Committee.

Some modification of projects for the anniversary celebration occurred as they developed. A brochure had been printed for the twenty-fifth anniversary; the initial plan was to update it for the fiftieth anniversary. As described by the head of the committee, Mary Lou Williamson, editor of the *News Review*, "the booklet will be a factual history of our city's origins and growth, its government and other public institutions, larger private institutions, its residential neighborhoods and homeowners associations, and its churches and civic associations."[17] The subcommittee sent letters to all local organizations asking them to write or to update their histories. In the next year, the plan for a brochure of compiled histories evolved into something much more elaborate. The goal became a pictorial history of Greenbelt, made possible by the collection of material in the Tugwell Room, back issues of the *News Review*, and pictures taken by the Farm Security Administration.

One subcommittee soon realized the need for expert help and created another committee to provide it. The subcommittee in charge of the conference on planned communities formed an Advisory Committee of "noted scholars in the fields of history, planning, housing and architecture."[18] This group met with the subcommittee to advise it on the shape and substance of the conference.

Throughout 1984–86 dozens of people worked to plan the year-long series of events held in 1987. The first of these occurred January 18, 1987, in the Utopia Theater, a kickoff celebration with reminiscences, songs of the 1930s, a slide show, and a preview of coming events. On February 15, at the twenty-sixth annual Washington's Birthday Marathon held at the NASA Recreation Center, all participants received T-shirts with Greenbelt's fiftieth anniversary logo. On February 18 and 25 Prince George's Community College and the Greenbelt branch of the Prince George's County Library sponsored a course on Greenbelt's history. On March 8 the American Legion offered a day of food and entertainment, with proceeds going to benefit the town's fiftieth anniversary celebration. On April 20 and 23 the local cable television station aired videotaped interviews with early residents.

The Greenbelt Conference on New Towns, held the weekend of May 1–3, began on Friday evening with the grand opening of the photography exhibit and a reception at the Greenbelt Hilton. The exhibit featured forty-two photographs taken by Farm Security Administration photographers such as Marion Post Wolcott, Arthur Rothstein, Marjory Collins, and Gretchen Van Tassel. The conference itself began Saturday morning with a keynote address, "The Vision of the New Deal," by the historian William Leuchtenburg. Other sessions included "Greenbelt—Implementing the Vision," "Reflections on Making the Greentowns Work," "The Future of New Towns," and "New Towns Housing the World." All sessions may not have met university academic standards, but the conference was a remarkable venture by a group of citizen volunteers. A five-part series summarizing the conference sessions appeared in the *News Review,* thus reaching a wider audience than those in attendance at the Hilton.

In the concluding session of the conference, the architecture critic Frederick Gutheim observed that although foreign visitors remained interested in Greenbelt "Americans have not paid enough attention to the Greenbelt effort." David Graham Hall, president of the Habitat International Council, a United Nations subcommittee, hoped that the conference would initiate a "new awakening of urban development in America."[19] The May 7 *News Review* summarized resident reactions to the conference in an editorial:

> We Greenbelters like to tell ourselves how wonderful we are. We also like to share that message with others. But over the last weekend at the remarkable international Greenbelt Conference on New Towns, we had the great and continuing pleasure of hearing notable architects and urban planners from around the nation and abroad

confirm what we already knew: Greenbelt is Great. . . . In the elegant ambiance of the Greenbelt Hilton, the town and its citizens were on view at their dressed-up very best—not a tee shirt in sight. The legend we have all worn so many times spread across our chests was invisible during the entire weekend. But even the outsiders knew that we carried it in our hearts.

The next major event of the anniversary year, house and garden tours, took place just two weeks later, on the weekend of May 16 and 17. More than sixty houses of all ages and types and from all sections of town were open for tours. On May 17 a special Anniversary Garden was dedicated in Greenbriar. On June 2, Nostalgia Night occurred at the Greenbelt Hilton ballroom. This event, held to raise funds for the museum, featured folksinger Joe Glazer singing songs of the New Deal era. William Lassell, an original occupant of 10-B Crescent Road, now the museum, recounted his memories of moving-in day in 1937. Greenbelt Day in June featured two special events. On June 6, at noon, about a thousand Greenbelters proudly gathered to have their picture taken at Braden Field. This picture appears on the dedication page of *Greenbelt: History of a New Town*, the pictorial history. Greenbelt Day at NASA attracted 2,000 visitors to the test facilities and the Visitor Center, transported from Roosevelt Center on a double-decker bus.

The 1987 Labor Day festival parade featured a float celebrating Greenbelt's fiftieth anniversary as well as the two hundredth anniversary of the United States Constitution displaying a replica of the "We the People" panel of Lenore Thomas's relief carved on Center School. The float proclaimed, "We the People . . . make Greenbelt and America Great!" Three generations of the Barcus family, fifty-year residents representing Greenbelt's pioneer citizens, rode on the float. Homecoming Weekend, October 9–11, welcomed past residents to town for the reunion. Weekend highlights included the dedication of the museum, a dinner-dance at the Hilton ballroom, and a luncheon for pioneer families sponsored by the American Legion. While the Conference on New Towns focused on life in Greenbelt in an intellectual, analytic manner, Homecoming Weekend celebrated the emotional ties among Greenbelters. In November the *News Review* celebrated its fiftieth anniversary with a celebratory dinner and a special fiftieth anniversary edition restating the paper's original seven goals first published on November 24, 1937.

Virtually all groups active in town life took part in the festivities. In addition to the special activities, the Fiftieth Anniversary Committee arranged with the Greenbelt post office for a special cancellation: "50th Anniversary, Greenbelt, Maryland, 1937–1987." The NASA stamp club produced a first-day cover commemorating the fiftieth anniversary of Greenbelt's charter on June 1, 1937. Citizens for Greenbelt produced a new map of the city. Greenbelt residents also created three long-lasting memorials of their fifty-year story: their museum and two books. *Greenbelt: History of a New Town, 1937–1987* is a pictorial history with a summary of each period of town life as well as brief descriptions of Greenbelt activities and organizations. *Greenbelt Is 50, 1937–1987: Looking Back* contains a compilation of the oral histories conducted by Dorothy Lauber, University of Maryland students, and the Oral History Subcommittee. The museum and both books preserve in tangible form much of what people celebrate by their slogan, "Greenbelt is Great!"

The fiftieth anniversary celebration garnered much attention for the city, but such notice had been fairly constant throughout the town's history. This attention served as a continual reminder to Greenbelters that they lived in a special place, thus helping them maintain their belief in their unique identity. After carrying out its successful rehabilitation project, Greenbelt Homes received even more visitors than usual from members of co-ops around the world, a typical example being a group from the Special Committee of the Council on Building Restoration in Sweden, who spent a day in town. A June 10, 1982, *News Review* article about the touring officials pointed out that 15 percent of all housing in Sweden is cooperative. A group of twenty-nine mayors of cities in Iran, touring the United States under the auspices of the U.S. Agency for International Development, visited Greenbelt to witness local democracy in action.

Area journalists continued to study Greenbelt's progress over time. The *Washingtonian* examined Greenbelt for its article, "How the New Deal Changed Washington." The *Washington Post* featured the town in two sympathetic articles, "A 'Green Town' Built in Thirties Battles to Stay Model Community," and "Greenbelt: U.S. Dream Town." According to the *Post,* Greenbelt "remains a community rich in green space and a town where moderate-income housing can still be found. . . . Today, Greenbelt residents speak highly of the town, proud of its planned past and proud that it is not the typical suburban community. The plentiful trees and greenery and the unusual architecture serve as tangible evidence of the original experiment."[20]

Prince George's County newspapers frequently weighed in with their opinions. An August 10, 1978, editorial in the *Prince George's Post* compared Greenbelt with Bowie:

When Franklin D. Roosevelt's New Deal planners designed old Greenbelt in the 1930s, they knew what they were doing. The ease of life in that comprehensively planned little town is notorious. Senior citizens and six-year-olds are equally comfortable in a community where residential streets radiate from a town center that includes police, library, grocery store, elementary school and parks. Architects managed to give both apartment houses and homes five-minute access to that town center. . . . The newer parts of Bowie offer a stark contrast to life in Greenbelt.

Bowie is described as having "slipshod workmanship," with much gasoline and time expended to reach shopping areas along Route 450. The editorial voiced a plea to those who plan communities to "think of people—we the people—in your designs, just as FDR's enlightened planners did in creating Greenbelt." A *Prince George's Sentinel* article posed a question: "Greenbelt: 'Ruined City' or 'Still the Best Place to Live'?" The *Prince George's Journal* featured Greenbelt in its article, "Model City Enters the 1980's."[21] All of these articles centered on Greenbelt's origins or on the differences between Greenbelt and other towns, clarifying exactly what Greenbelters had fought so hard not to become in the 1960s, just another suburb, having all its unique features obliterated by "development." In the early 1980s others began to appreciate the vision held by Greenbelt residents as they retained their Greenbelt principles.

In November 1984 a television crew spent a day in town, filming scenes for a one-hour documentary on Franklin Roosevelt's "lasting contribution as President of the United States." In June 1985 a *New York Times* Sunday edition featured an article on the town's preparations for its fiftieth anniversary.[22] In November 1986 city planners from Korea and Iraq visited Greenbelt as part of a UN-sponsored global tour of "new cities." They met with Greenbelt's planning and development coordinator and toured the town. While in the area they also visited the new towns of Reston, Virginia, and Columbia, Maryland.

Whatever characteristics Greenbelters have, modesty is not one of them, and a large part of their fiftieth anniversary celebration involved letting the outside world know what they were celebrating and why. Greenbelt residents succeeded in this endeavor, as their anniversary received a great deal of attention. Television, newspapers, magazines, and academic journals all paid attention to Greenbelt's fifty-year history. WETA, the public television station in Washington, D.C., produced a documentary on Greenbelt, Reston, and Columbia called *The New Towns*, aired on May 1, 1987. The May 4, 1987, *Washington Post* ran a lengthy article on Greenbelt's anniversary headlined "Greenbelt Survives Development with Its New Deal Spirit Intact," pointing

Greenbelt's "first families" as guests of honor at dedication
of Greenbelt Museum, October 10, 1987

out that "Greenbelt residents led the way in opposing unplanned development in the 1960s when the county was approving large-scale apartments. Since then, the Greenbelt vision of quality development has become the theme of the current county administration. 'Long before the county government started being aggressive in demanding quality growth, Greenbelt was adamant,' said County Executive Parris Glendening." Greenbelt had thus come full circle. By insisting on retaining the New Deal ideals of its past, the city acted as a harbinger for the future.

On the floor of the United States Senate, Maryland Senator Paul Sarbanes spoke of the anniversary: "Mr. President, I join Greenbelt's citizens in celebration of its fiftieth anniversary. The deep devotion of Greenbelt's residents to their community is proof that the experiment has worked. I ask unanimous consent that a recent article from the *Washington Post* describing this fine city be included in the *Congressional Record*."[23] City planners took special note of the anniversary, with articles on Greenbelt appearing in *Landscape Architecture,* the *American Planning Association Journal,* and *Urban Land.* Lloyd Bookout, the author of the *Urban Land* article, concluded:

> Did Greenbelt, Maryland, become the viable community envisioned by New Dealers? Conference attendees overwhelmingly agreed that it did. The community itself worked together to preserve and enhance the dream of the town's founders. The spirit of the community could not be quelled as easily as the program that started it. Although circumstances have resulted in changes to the original physical plan, Greenbelt is still a community characterized by extraordinary citizen involvement and pride—not a mean accomplishment. In sponsoring this conference and the other anniversary festivities, the citizens of Greenbelt are sending out the message that, for them, Greenbelt is an experiment that worked.[24]

At the American Planning Association's national conference in April 1987, Greenbelt received special recognition. The National Historical Landmarks program of the American Institute of Certified Planners honored greenbelt towns for the significant contributions of their design principles.[25]

While the special attention of outsiders helped residents appreciate their uniqueness, one local institution, the *News Review,* kept as its prime mission the persistence of the Greenbelt idea. As is obvious from the many mentions here of the paper's role, the importance of the community newspaper in the maintenance of Greenbelt's original goals and ideology can hardly be overstated. The paper reminded readers of town history and traditions and provided both communication among the increasingly far-flung residents and information on significant current events.

Even during the tremendous battles of the 1960s the newspaper staff continued its normal community functions. They covered local politics in an effort to keep readers informed, especially important in the time of pro- and anti-development factions. For each such election, the paper published biographies of all candidates and discussed the issues from all sides. A new era in local politics began with the purchase by candidates of large advertisements extolling their merits. When covering controversial issues, the staff solicited candidates opinions and published their views. Before each city council meeting the paper published the agenda on page 1; later it reported the details of the meetings. In the continuous discussions on development the paper not only reported on all zoning hearings but also gave the history of the ownership of each land parcel in an attempt to put each decision in perspective.

The staff made special efforts to respond to the continuing growth of the city. New columns featured newly built sections of town, such as "Springhill Lake Notes" and "Boxwood Village Notes." In November 1964, during an influx of new residents, the paper ran an article on its own history for the benefit of newcomers. Again, on October 26, 1967, the paper introduced itself to new readers in the University Square apartments via an editorial:

> We do not know whether you realized, when you chose to live here in Greenbelt, that there was something special about this town. Oh, don't ask us to define it—it's a mutual identity, a cooperative spirit, an elusive elan, and it floats around, adding zing to council meetings, creating and sustaining countless projects and associations (as well as those tempests-in-a-teapot for which Greenbelt is known), giving us all a feeling of sharing the life of a community, to which we belong.
>
> Whether you bargained for all this when you came, or were merely weighing commuting times against rents and access to schools, we welcome you now as new residents of our town. Perhaps you have already discovered our tranquil lake with its parkland setting, and other features unique to this city. We now invite you to discover something no less characteristic—the Greenbelt News Review.

In 1970 the News Review staff celebrated its thirty-third anniversary with an open house as well as with some analysis and reflection, noting that by November 1970 there had been thirty-nine editors and that in 1970 twenty-two staff members had provided 4,500 free copies of the paper.

A questionnaire given out by the city council at the polls that November showed that 78 percent of the city population read the News Review. In Greenbelt Homes houses, 97 percent read it, while in the single-family homes, 95 percent did so. In Springhill Lake, only 44 percent said they read it, but due to distribution problems many residents did not receive it regularly. These figures reflect the level of interest in the goals and ideals of the original town in the various sectors of 1970 Greenbelt. In the Greenbelt Homes section, the original planned community, virtually everyone read the local news. In the single-family homes occupied by many former Greenbelt Homes residents, interest in the local community was roughly the same as in the original homes. In the new apartment area across the Beltway from the rest of town, less than half of the residents read the local news. Actually, it is surprising that as many as 44 percent read the *News Review,* as these residents dwelt in isolation from the rest of the community, with no inherent interest in the original goals and plans of the city.

The staff continually worked to improve distribution in new areas, in hopes of increasing readership. However, they admitted that lack of coverage of events in the newer parts of town continued to be a problem. In a July 25, 1970, editorial entitled "A Self Appraisal," they concluded:

> If there is an imbalance in coverage, the answer, of course, is not to cut down GHI [Greenbelt Homes] coverage, but to improve the coverage of news from other sections of town. To do so will require the cooperation of residents of the various areas, because there are few focal centers of information to which staff reporters can turn. In effect, we are asking for help, for volunteers in various sections to become special correspondents. No experience is needed for such volunteer work. Remember, this is your community newspaper. We gather every Tuesday evening, 8 to 10 P.M. at 15 Parkway (basement). Come down and get acquainted.

During periods of stress, which included most of the 1960s, Greenbelters vented their feelings in letters to the editor. At intervals, the staff reminded readers of pertinent facts. This one from March 14, 1968 is typical:

> Letters to the Editor represent only the personal views of the person who signs them. Publication by the *News Review* of these letters indicates neither approval nor disapproval of the views expressed. . . . It sometimes happens that in controversial issues most of the letters come from one side. Readers have complained to us that in such situations by printing all of them we have in effect endorsed the side that has written most of the letters. We do not see how we can do otherwise. All we can ask is understanding from our readers that the editorial policy and position of the *News Review* is expressed through its editorials and not through the Letters to the Editor Department.

On April 29, 1972, the *News Review* staff held a testimonial dinner to celebrate their victory in the Supreme Court suit. The dinner was attended by

175 Greenbelters, who heard the guest of honor, the attorney Roger Clark, talk about the case and its outcome. The *News Review* published a special issue the following November, on its thirty-fifth anniversary, honoring Elaine and Al Skolnik. Al had been president of the board since 1959, while Elaine, in her duties as reporter, became an expert on zoning. The burden of the court case weighed heavily on them, but events such as the testimonial dinner and being named Greenbelt Citizens of the Year in 1974 demonstrated that the community appreciated the sacrifices they had made.

The paper continued to emphasize town history at every opportunity. In 1971 a series of excerpts appeared from "Atlantis on the Hill: A History of Greenbelt, Maryland," by Robert Muller. In 1972 a series called "35 Years Ago" ran, as well as a special thirty-fifth anniversary issue. When new housing areas opened and newspaper delivery began there, the paper would greet readers with a brief history of Greenbelt and of the paper. In 1973 the paper published a series called "25 Years Ago." When Greenbelt Homes celebrated its twenty-fifth anniversary, the fire department and rescue squad its thirtieth anniversary, and the Labor Day festival its forty-second anniversary, articles detailed their histories. A feature on the twentieth anniversary of James Giese as city manager pointed out his quiet competence and able leadership in dealing with tumultuous city affairs. A focus on Greenbelt as a planned community led to an article titled "Roosevelt, New Jersey: A Visit to a Sister City" and to coverage of Giese's visit to Greenhills and editor Mary Lou Williamson's visit to Greendale. The article on Greenhills concluded, "Greenbelt makes history buffs of a lot of us."[26]

The newspaper staff attempted to maintain community interest in significant figures from Greenbelt's past. An article in May 1975, "Remembering Clarence Stein," appeared soon after his death, providing reminders of the part he played in Greenbelt's design. Franklin Roosevelt Jr.'s visit to town in October 1976 received detailed coverage. When Rexford Tugwell died on July 21, 1979, a front-page article in the August 2 issue recalled to readers his ideals, which had led directly to the formation of their town. The city council passed a resolution in Tugwell's honor, which stated in part: "Be it further resolved that this City and its citizens are honored to serve as one of many monuments to the life of Rexford Guy Tugwell and to be known as a 'Tugwell Town.' " The derisive label had become one of distinction. Grace Tugwell's visit to town in 1982 received front-page coverage.

As the years passed, the affairs of Greenbelt Homes, which had always been front-page news, began to be relegated to later pages, while other parts of town received more prominent coverage. The staff took an occasional break from serious issues, revealing its lighthearted side. Whenever the paper publication date fell on April 1, an April Fool issue appeared. The 1976 April 1 issue looked normal; however, the stories regarding current issues were just outrageous enough to provoke laughter and the question, Can this be true? The reader then turned the page to discover the "real" front page underneath.

The paper increased its efforts to unite the separate parts of the city throughout the 1970s. Typical is the January 26, 1978, editorial, "Invitation to Greenbriar": "Two blocks from the Center! That is the approximate distance between residents of the Greenbriar development and the heartbeat of this town to which they pay their taxes and which invites them to participate in its ongoing concerns and enthusiasms, to accept its services and to feel the invigorating warmth of its sense of community." The Baltimore-Washington Parkway functioned as "a veritable Grand Canyon in its restriction of communication between these two parts of the city." The article pointed out the advantages of the pedestrian overpass for Greenbriar residents, who could come to the Center on foot, and "be a Greenbelter."

During the 1970s, as the pioneer residents aged, the *News Review* reported on the retirement or death of a number of individuals who had been important to the town in the early years. These articles reminded old-timers and informed newcomers of the city's history and the significant contributions that one individual could make. Front-page articles featured the death of Allen Morrison, Greenbelt's mayor during World War II; the retirement and death of Buddy Attick, who helped construct the town and then remained, becoming director of public works; and Robert McGee, a community activist. The most painful to report was the sudden death of Al Skolnik, aged fifty-six, of a heart attack. The March 17, 1977, *News Review* headlined: "Greenbelt Stunned by Untimely Death of Alfred M. Skolnik." An editorial listed his numerous contributions to the paper:

> Perhaps most important of all to the newspaper was Al's role as mediator. He handled with courtesy and quiet firmness the importunities of overexcited Greenbelters demanding that the paper adopt their view of an issue as our editorial stance. He led the staff in coming to consensus on an issue for an editorial position, or, if the staff were split, to avoid taking a position. Under his influence over the years, the disparate personalities of the staff have moved together in harmony. He respected us, and we respected him. He made us stretch to our utmost achievement. His solid encouragement helped many of us to attempt and succeed in tasks

we would never have dreamed we could accomplish—and for the enhancement this brought to our own lives, we owe a debt to Al.

As a longtime friend described it, "Al was the conscience of the community." The same issue of the paper contained tributes to Al Skolnik from many in town, as well as a city council resolution, which concluded, "We commend as a model to all the integrity, selflessness, sense of fairness and dedication to service to the public displayed by Alfred M. Skolnik." On April 14 another editorial reflected: "In the past three months, this community has suffered three grievous losses: Edgar Smith, former mayor; Alfred Skolnik, president of the *News Review;* and now Bob McGee, an outstanding citizen. We know these people will be missed and that we will be unable to replace them. Yet their legacy to us is the inspiration to continue their efforts, to work always to improve our family—our special Greenbelt family." In April 1977 Elaine Skolnik was elected president of the board of the *News Review.* In May 1978 she presented the first Alfred M. Skolnik Memorial Award to a graduate of Eleanor Roosevelt High School interested in a career in journalism.

Maintaining free weekly publication by an all-volunteer staff continued to be difficult. Beginning in 1978 the addition of journalism interns from the University of Maryland helped the staff to cope. By 1980 the number of Greenbelt volunteers dwindled, with too heavy a burden falling on too few people, reflecting the increasing number of households with both parents working and neither having free time for volunteer activities. The city manager Jim Giese wrote a letter to the editor in the May 8, 1980, issue urging additional people to get involved: "One of the main reasons for Greenbelt's being a unique and close community is that it has its own citizen-operated weekly newspaper. It is too important to our community to be treated lightly or ignored." That Giese's appeal produced results is revealed in the fiftieth anniversary issue of December 31, 1987, in which *News Review* volunteers explained how they got involved. One volunteer wrote:

My career with the *News Review* started almost seven years ago as the result of a simple act: City Manager James K. Giese wrote an open letter to the paper's readers pointing out that the community could not expect the same people who had devoted so many years to the paper to do it forever and warning that we would be in danger of losing this valuable community resource if more residents did not volunteer. That was enough for me. I valued the paper and regarded it as an important element in the glue which holds this community together. So I called Elaine Skolnik and said that I would write a few business letters each week or do some proofreading. I've barely had a day off since!

Another staffer remembered, "I started working on the advertising desk and have since gotten involved in many aspects of the paper. Two years ago I became the third of the five-member board who originally joined the paper in response to the city manager's letter." In January 1981 an editorial asked more people in the community to become "stringers," "a part-time representative for a news publication." Items of news went unreported because the small staff could not deal with everything of interest. A new problem developed in 1982: the number of children in town had decreased and there were too few available to deliver the paper. A plea for help suggested that mothers with young children, retirees, or people who simply liked to walk could help deliver the paper.

During the years preceding the fiftieth anniversary celebration in 1987, the *News Review* did its part to pave the way. From July to September 1984 the paper ran a five-part series on the history of Greenbelt. Beginning in January 1985, monthly articles with the "Greenbelt Is 50" logo appeared, along with a feature called "Fifty Years Ago." As always, the newspaper did its best to help faltering community institutions. An editorial on February 23, 1984, was titled, "Utopia Theatre—Here to Stay?" The editorial concluded, "Two outstanding theatrical productions went practically unnoticed by a substantial number of Greenbelters. It was a great loss. But the greater loss may be at our doorstep. Supporting the Utopia Theater can only bring us rewards—in entertainment and community pride."

When the newspaper staff felt it necessary, they editorialized on town issues. A typical example from July 10, 1986, decries the dissension in the city council: "We do not doubt that all members of council are deeply devoted to Greenbelt's interests. All have proven their devotion in years of service to the city, with all the sacrifices of time, energy, and private pursuits that such service requires. All in Greenbelt must appreciate those contributions. But the inability of council members to resolve their personal differences has gotten out of hand. We think they can do better." Reflecting the changing nature of Greenbelt, a new feature made its appearance in 1986, "Greenbelt's Business," a page about business activities and people.

In May 1986 the Prince George's county executive Parris Glendening presented an award to the *News Review,* which was selected as one of the top ten volunteer organizations in the county that year. The award honored the paper for "helping to maintain a strong community spirit in the city of Greenbelt and for the commitment and the dedication of its staff."[27] However, the paper so honored continued to have a difficult time surviving due both to a financial squeeze and to decreasing numbers of volunteers. At a special meeting on March 11, 1985, board president Elaine Skolnik spoke to more than twenty-five representatives of the city's organizations:

> Our small staff is so busy publishing the paper each week that we cannot address the problems of finding volunteers and of soliciting more advertising. If the *News Review* is to see its 50th year—and the years beyond that—and if Greenbelters are to continue to know what goes on in their city, the *News Review*—we—desperately need your help. We must find new blood—a new generation of volunteers and we need to know we have replacements.[28]

On March 20, 1985, six people formed a committee, the Friends of the Greenbelt *News Review,* and met weekly thereafter, seeking both funds and volunteers. Committee members sent 500 letters to businesses and advertisers in the Greenbelt area, as well as 8,000 packets to homes in town. By July 25, the paper had received $10,009 in contributions from 511 donors, and 79 people had volunteered for work.

With the newspaper on a firmer footing, Elaine Skolnik relinquished her role as president of the board but continued to serve as news editor. She had served on the paper for thirty-one years at this point, having begun with the "Our Neighbors" column in 1952. Thus the newspaper entered its fiftieth year with both old-timers and new workers, as well as more financial security than usual. The improvement in the newspaper's situation boded well for the celebration of Greenbelt's fiftieth anniversary, as the *News Review* provided the necessary communication among townspeople to plan and carry out the massive celebration.

If any one community institution proved crucial to the persistence of Greenbelt's original goals and ideals, it is surely the *News Review.* The newspaper continues to serve in several key roles: providing communication among townspeople, functioning as the chief mechanism by which organizations are created and maintained, and educating residents on the town's past, especially important for the continuous stream of newcomers in new areas. Greenbelt has maintained its identity as a cooperative, planned community due to its ability to keep its history alive in the present.

# 8

# Greenbelt Today
# and Tomorrow

AFTER RECOVERING from the exertions of their fiftieth anniversary celebration, Greenbelters worked their way toward the millennium in their usual fashion, maintaining their traditions, organizations, and cooperatives. Fiftieth anniversaries continued to be celebrated by community groups and organizations. Reminders of the past continued to be emphasized. At the 1988 Labor Day festival, the presenter of the Outstanding Citizen of the Year award commented:

> A year ago, this historic city was celebrating its 50th anniversary. . . . The celebration has officially ended, but the history lessons we relearned have carried over. We have, for example, been reminded of the people who built some of our most important institutions, and sustained them through periods of crisis and change. The 1988 outstanding citizen, Bruce Bowman, has been one of those institution builders. This person's achievements have been historic—and they are still important today.[1]

Labor Day festival celebrants also unveiled a plaque marking the first home occupied in Greenbelt, 1G Gardenway, whose residents moved in on September 30, 1937. Also in September 1988 Prince George's Community College and the Prince George's County Library system sponsored a two-session class on the history of Greenbelt, which included a walking tour and a visit to the museum. The idea remained alive for a proposed memorial for

Franklin Roosevelt, brought up for the fiftieth anniversary celebration but not carried out, as the Friends of the Greenbelt Museum sponsored a program on the topic in February 1989.

The city continued its own development, building a new police station near the intersection of Kenilworth and Crescent Roads, in art deco style to match earlier city buildings. The groundbreaking for the indoor pool occurred on September 6, 1990, with its neo art deco architecture blending well with the design of the original, adjacent, outdoor pool. The dedication was held one year later, after which Mayor Gil Weidenfeld was ceremoniously tossed into the pool. The city gained another building in 1995, when a federal courthouse was built at the instigation of Representative Steny Hoyer. It was located adjacent to Capital Office Park.

One Greenbelt institution gained the spotlight unexpectedly in 1989. An innovative program at Center School illustrates the new types of problem that were brought to town along with the pupils imported by busing. Principal John Van Schoonhoven heard reports from concerned teachers of children from the Washington Heights neighborhood who feared returning home after school. Drug dealers reigned in their neighborhood, roaming the streets with impunity. Some children lacked even minimally adequate food and clothing. Van Schoonhoven met with the county executive Parris Glendening, the school superintendent John Murphy, the police chief Michael Flaherty, and school counselors and administrators to fund an after-school and summer program for these children and any others who wished to participate. The children in the program play outside, get assistance with homework, go on field trips, and participate in sports.

When the *Washington Post* ran the series, "Victims of Drugs," featuring a child living in a crack house, it was brought up by President George Bush at a news conference. For Dooney Waters, "school was the only haven of hope."

> Field [Dooney's teacher] supplied much of Dooney's wardrobe, sent food home in his book bag each night, and arranged for families in Dooney's neighborhood to cook him hot meals during the winter months. Field also devised a plan for Dooney to receive extra food through the school's free lunch program. Denied regular meals in his home, Dooney often gobbled down two meals at lunchtime and sometimes scrounged around for leftover items on other pupils' trays. Joe Squarells, a custodian at Roosevelt Center, visited Dooney's apartment on weekends and sometimes took him along on family outings. Teacher Marcia Donahoe Marino bought Dooney a bag of athletic socks and a pair of high-top tennis shoes in late May. His old sneakers, the same ones that he had when he started kindergarten in 1987, were so small that his toes had begun to bruise and blister.[2]

School officials tried to get county social service workers involved in Dooney's plight without much success until the story made the front page of the *Post*. As the *Post* summed it up in an editorial, "In Dooney's case, the school responded in a way that was above and beyond its responsibility. The county government fell short."[3] The school's special programs continue, with the principal Van Schoonhoven encouraging volunteer help from the community.

The *News Review,* with a volunteer staff of seventy-three, continued to serve the community in its usual manner, especially by recalling to readers the city's history at every opportunity. An article on Lenore Thomas Straus at the time of her death in January 1988 reminded readers that she had created the friezes on Center School as well as the statue, *Mother and Child,* at the Roosevelt Center. An article on the death of Tilford Dudley in early 1990 recalled for readers his participation in land acquisition for the greenbelt towns program.

The *News Review* has changed gradually over the years. Two early functions of the paper have been successfully taken over by the community. First, in the early years, the paper staff had to continually remind residents to observe various holidays and traditions, but volunteers in the community now maintain these without urging from the paper. Second, for many years, editorials pointed out problems that needed to be addressed, but since the 1960s residents have seen problems for themselves and taken action. Often the first action involves the paper, with a notice or a letter to the editor to alert fellow residents. Although a community resident initiates the action, the paper functions as the preferred method of communication.

Another change in the paper, a decrease in editorials, especially controversial ones, dates from the era of the Supreme Court suit. Some deny this; others agree that change has indeed occurred and that it is in reaction to the Supreme Court suit.[4] Even though the *News Review* won the case, it proved a long, harsh experience, and the newspaper staff may, consciously or otherwise, try to make sure that it is not repeated.

The *News Review* currently distributes 10,500 free copies of the paper weekly to homes in central Greenbelt, Greenbelt East, and Springhill Lake and to office buildings and shopping centers. Over the years the paper has had thirty-nine editors; however, Mary Lou Williamson has held that position for the past twenty-four years. The newspaper staff continues to be augmented by student interns from the Departments of English and Journalism at the University of Maryland, an arrangement thought beneficial to all concerned.[5]

While Greenbelters maintain their old co-ops, such as the newspaper and grocery store, they also form new co-ops to meet their evolving needs. The April 11, 1996, *News Review* reported on the latest, GREENBELT.COM, a group that planned "to make the Internet available at wholesale cost to Greenbelters of all kinds: that is, businesses, individuals, game players, workers, and learners." Beginning their organizational work in April 1996, they have succeeded in making Greenbelt a "Web-connected city."

In addition to co-ops, issues of development and land usage continue to be vitally important to Greenbelters. In the aftermath of their fiftieth anniversary celebration, many town residents rededicated themselves to reclaiming as much of their greenbelt as possible. In 1990 the county executive and council stressed the importance of enforcing zoning limits and encouraging the preservation of green space. These beliefs echoed those of Greenbelt residents in the early 1960s, when extensive development began.

The continuing reclamation of pieces of the greenbelt is a time-consuming and expensive process, involving much dispute and not only from expected sources. The owner of parcels 1 and 2, Charles Bresler, stated that the city had plenty of green space and needed more "upscale housing."[6] In 1989 the city council favored "low density development" for these parcels of land but experienced a change of mind when lobbied by the Committee to Save the Green Belt.[7] In addition to mass attendance at council meetings, booths at Labor Day festivals, letters to the *News Review,* letters to local, state, and national politicians, the committee conducted guided tours of the untouched forestland each Saturday so people could view for themselves the areas in dispute. Their activities generated much support, in spite of the fact that most residents agreed on the desirability of building more single-family homes in town. Many Greenbelters felt that the preservation of the greenbelt took precedence even over the wished-for homes.

However, the goal to preserve original forest tracts did not meet with approval from everyone. Many in Greenbelt East saw no need to spend millions of dollars of city money for this purpose. The Greenbelt East Advisory Committee chairman stated, "There's a Greenbelt—I don't want to call it a myth—but in the original Greenbelt self-image, people think of it as a small town surrounded by green space with the town center as the hub of activity. They know we're here but they'd rather not think of us. It destroys an idea, an image. So I think they shut us out of their minds."[8] The speaker accurately separated the Greenbelt ideology of the past from the new physical reality in a way that many in Old Greenbelt would rather not face.

Greenbelt East lacked the clout necessary to stop the Committee to Save the Greenbelt because its residents were not as politically active as those in Old Greenbelt; thus the city council decided to purchase parcel 2 from Bresler for $360,000 in October 1988. In June 1989 the county school board gave its surplus school land adjacent to parcels 1 and 2 to the city. The county also obtained state funds to help the city buy parcel 1. By June 1990, after much negotiation, the state's woodland conservation funds furnished $1,250,000, and thanks to the assistance of U.S. Representative Steny Hoyer, the National Park Service contributed $500,000 for partial ownership of parcel 2. The willingness of state and national officials to be involved in this effort demonstrated that it was not only Greenbelters who saw the importance of saving green space.

The acquisition of this land did not prevent disputes regarding its use. The elderly utilized their considerable political power to ask that some of the undeveloped land be used for a 100-unit cooperative apartment building for senior citizens. Their wishes fell before those of the majority, who insisted that the land remain parkland and who felt it was not right to use city land solely for the benefit of one group.[9] The city itself proposed to build a storage compound on part of the land, compelling a new resident to write to the paper:

> On June 4 I attended my first Greenbelt Public Hearing. I enjoyed the community flavor and the intensity of the topic. But even more, I was overwhelmed by the passionate concerns of the citizens. . . . It seems to me that Monday's meeting was not simply about a storage compound. It was a gathering that demonstrated what has long been a history of Greenbelters intervening to save their forestland.[10]

The writer urged the creation of a permanent wildlife preserve. Residents convinced the city to build nothing on the land in question.

Members of the Committee to Save the Greenbelt and other interested individuals carefully monitored the Washington Metropolitan Area Transit Authority's activities involving construction of a Metro subway station in Greenbelt. The groundbreaking ceremony for the station occurred in October 1989. Most Greenbelters agreed that the subway would be beneficial, but some expressed concern about the adjacent train yard planned on Beltsville Agricultural Research Center (BARC) property. The Metro purchased seventy acres of land from the agricultural center for the train yard, but many environmental activists in Greenbelt maintained that these valuable wetlands should not be disturbed. Instead, they encouraged the utilization of a nearby property used for sand and gravel operations.[11] The committee purchased a large advertisement in the February 8, 1990, issue of the *News Review* to state their views and to encourage readers to write to Metro and county officials.

Greenbelt residents went into action again when, in 1996, they learned of the plans of the U.S. Department of Agriculture (USDA) to build an office complex on BARC land. The *Washington Post* front-page headline was "Greenbelt Residents Jump to Defend Its Precious Belt."[12] Residents met frequently with officials of the USDA; they were concerned with the loss of trees and green space on BARC land as well as with a massive influx of traffic on already overcrowded roads. Even though they lost the battle, Greenbelt residents continue to monitor the use of BARC property, realizing that it constitutes an important part of their greenbelt.

Greenbelters attempt to preserve not only their greenbelt but also the original Greenbelt buildings. By far the largest number of these compose Greenbelt Homes, Incorporated, which still strives to provide high-quality, affordable housing to its members. A decision on whether to ask Prince George's County for historic district designation for Greenbelt currently faces the city council. First proposed in 1993, the idea resulted in a 1994 study carried out by Howard Berger, an architectural historian and preservation planner for the Maryland–National Capital Park and Planning Commission.[13] Those attending the April 1996 meeting of Citizens for Greenbelt discussed the issue at length. James Giese, the former city manager, pointed out that the importance of Greenbelt lies in its plan rather than in its buildings, which can be aptly described as "blocks with windows and doors." Residents worry that historic designation would make changing their homes difficult. As one pointed out, "I agree, a Historic District is great, but we've got to live here. Williamsburg is great, but nobody lives there."[14] No one wishes to ossify Greenbelt exactly as it appeared in 1937–41, but the question of what should be preserved and the mechanism for doing so will continue to vex Greenbelt Homes members. In spite of doubts, they voted two to one in favor of historic district designation in April 1999. The city council has not yet taken the action necessary to complete the process, as questions remain regarding the effect of historic designation.[15]

In July 1983 Greenbelters received a distinctly unpleasant surprise when

Springhill Lake development (interior courtyards and
walkways are similar to those of original Greenbelt).
Photo courtesy Greenbelt Museum

the county board of education announced plans to demolish Center School
and replace it with a larger facility. The fact that the school, along with all of
the original town, was listed on the National Historic Register did not seem to
disturb the board greatly. In August, at a public hearing, historians, politicians, community leaders, and concerned residents spoke out against the
plan, arguing that the school served as the focal point of the original planned
town and was regarded as one of the most distinguished art deco buildings in
the nation. On August 18 the school board announced its unanimous decision to modernize rather than demolish the school.[16] In 1984 the Prince
George's County councilman Richard Castaldi, who recalled climbing on the
art deco sculptures of Center School as a boy, drafted a bill to place the school
on the county's Historic Sites and District Plan. The passage of the bill
provided protection for the structure.

During this period the fate of North End School also received consideration. After its closure by the county board of education, the school site

engaged the attention of numerous private interests. However, a recently passed state law gave the first option of purchase to the city, which quickly bought it. In July 1986 the Citizens for North End Center formed to develop plans for the use of the facility. Five years after the city purchased the site, with several studies complete but no concrete plans developed, town residents began to agitate for the formation of definite proposals.

At this juncture, in January 1989, the county school board abruptly announced that the town could choose whether to renovate Center School or build a new school at the site of the old North End School. Two weeks after the announcement more than a hundred residents appeared at a public meeting to discuss the issue. If Center School closed as a school, it would be remodeled as a community center. Of course it had always functioned as a community center, but if a new school replaced it in the North End, then its function as a school would be lost. If the community chose to renovate Center School to maintain it as a school, then the old North End School would be remodeled into a community center.

The manner in which Greenbelters handled this issue illustrates perfectly the community's decisionmaking process. Letters appeared in a steady stream in the *News Review,* representing all possible points of view. Two groups formed, the Citizens for Education and Community, who wanted Center School restored as a community center with a new school in the North End, and the Citizens to Save Center School, who wished to keep Center School as a school and use the old North End school for a community center. The two groups wrote letters, took out full-page advertisements in the *News Review,* debated at public meetings, and distributed flyers. Finally, as opinion appeared evenly divided, the city council decided to hold a community vote.

City voters expressed their wishes in a referendum in June 1989 in the form of a bond issue. A $3 million bond ordinance for a community center received narrow approval, 52–48 percent. An even narrower margin of voters selected Center School as the new site for the community center with 50.3 percent voting for it. An additional $1 million bond ordinance for the development of Schrom Hills Park in Greenbelt East passed easily, with 75 percent of the votes cast.[17] City officials dedicated the park on Greenbelt Day, Sunday, June 7, 1992.

For years the Community Center Task Force, made up of town residents, city staff, architects, and contractors, labored to turn Center School into the Greenbelt Community Center. On March 16, 1996, the community celebrated its completion at a dedication and open house. The ceremony began outside, beneath the "We the People" frieze at the south entrance to the building, where Mayor Antoinette Bram cut a large green ribbon.[18]

After the ceremony moved inside to the auditorium, following numerous speeches of appreciation, the former city manager James Giese reminisced about events that had taken place in the room:

> This building is Greenbelt history and, today, you have become a part of it. This is where most of the churches of Greenbelt were formed and held their first worship services. This is where recreational programs were first organized and the Greenbelt Recreation Department became the first such department in Maryland. Here the Greenbelt library was organized—one of the first, if not the first, of the county's public libraries. . . .
>
> In this room Greenbelt's citizens organized a cooperative to run the stores in the commercial center, the predecessor of our present grocery store. This is where in 1953, the residents formed another cooperative, GHI, to buy the homes built by the federal government.
>
> This is the site of many occasions in the life of the city, both happy and sad. Eleanor Roosevelt came here, a fur stole draped about her shoulders, to support a war bond rally in 1942. Here the town held a memorial service when President Franklin D. Roosevelt died, and shortly thereafter celebrated the end of World War II. And on all important occasions the Greenbelt band played, as it does today.
>
> I remember seeing this room filled to standing room only in 1964 as more than a thousand citizens turned out to protest a proposed high-density Greenbelt master plan. GHI residents have often filled the room at annual meetings, expressing their divergent views on how best to run the cooperative.[19]

After giving a brief history of the dispute that resulted in the restored community center, Giese concluded:

> This building which we dedicate today will join our other fine community facilities, the new school, the Aquatic and Fitness Center, the Greenbelt branch of the library, the adjacent athletic fields and the Youth Center, as places where our citizens can come together, work together and have fun together. With these facilities and the many citizen organizations that use them, we carry on the ideals of the planners who first created Greenbelt. We continue to make our community an important part of our life, and we continue to make Greenbelt, Greenbelt.

Those in charge of the restoration attempted to produce a building looking as much like the 1937 original as possible, while meeting the current needs of the community. As manager Cathy Salgado commented: "Even the bathrooms were restored with period tile and color schemes."[20] The building now houses the *News Review* office, a multimedia room, a kitchen and dining hall, a senior game room, a senior lounge, rooms for child and adult art classes and for child and adult day care, a game room, the Greenbelt Museum, studios for art, ceramics, photography, and cable television, and the original auditorium-gymnasium. The New Deal Cafe, a co-op of course, operates on weekend evenings in the kitchen and dining room area, providing desserts and live music.

Greenbelters achieved a long-hoped-for distinction when, on February 18, 1997, Secretary of the Interior Bruce Babbitt designated Greenbelt a national historic landmark, the highest form of historic recognition granted by the federal government. Greenbelt is the first planned community to receive this honor. The designation came about, as usual, as the result of communal and committee effort. When Greenbelt Homes sought money for rehabilitation, the board of directors considered the usefulness of having Old Greenbelt designated a historic district by the county. Though this has not occurred, the Architectural and Environment Committee of Greenbelt Homes urged its board of directors to not discard the issue completely. In 1994 the board established a study committee, chaired by James Maher and John Downs, which recommended that the city and Greenbelt Homes pursue landmark status. Carolyn Pitts, a historian for the National Park Service, assisted in guiding the city and Greenbelt Homes through the lengthy review process.

Greenbelters pursued landmark status for two reasons: to raise awareness of their town and its history and to obtain the protection provided for federal undertakings, such as highway projects. To be designated as appropriate to receive national historic landmark status, the property in question must meet at least one of six criteria; Greenbelt met three. One is that the property be associated with a broad pattern of national history. In Greenbelt's case, the mass migration in the 1920s and 1930s of farmers from countryside to city created an urgent need for good housing. Federal officials designed Greenbelt as a direct response to this need by providing affordable housing in a suburb for the urban poor.

Greenbelt also met the criterion of a property associated with a great idea or ideal of American history. Greenbelt clearly demonstrates the American ideal of safe, healthy, affordable housing for all citizens, with the special goal of nurturing children in the community. Third, Greenbelt embodies distinguishing characteristics of an architectural type, being a planned community designed along garden city lines with outstanding examples of art deco and streamlined moderne architecture. Greenbelters proudly installed the plaque announcing their new status during the sixtieth anniversary celebrations.[21]

Even while Greenbelters celebrate the continuation of their ideals, most recently expressed by the remodeling of the community center, they are surrounded by modern-day suburbia. Greenbelt is now a city of numerous residential areas as well as shopping centers and office parks, cut into eight pieces by freeways and major roadways.

The town consists of three main residential areas, central Greenbelt, Springhill Lake, and Greenbelt East. The core, Old Greenbelt, remains the largest unbroken piece, and contains housing in addition to the original units. Located most centrally are the original housing, "defense" housing, and Greenbelt Plaza Apartments. Lakewood, Boxwood Village, and Woodland Hills, all single-family homes, and the adjacent apartment complex of Lakeside North, are north of the lake. Custom-built houses are currently appearing in a small area on Ridge Road north of Lakewood and on Research Road north of Ridge Road. South of Attick Lake Park and the Roosevelt Center are Charlestowne Village, Charlestowne North, and University Square apartments. An area of single-family homes known as Lakeside is directly east of the lake.

These housing areas form a united grouping between freeway barriers and compose the physical center of town. People in Greenbelt readily agree that residents in this central area identify themselves much more strongly with the city and the Greenbelt principles of planned, cooperative living than those east of the Baltimore-Washington Parkway or west of the Beltway. John Buckner, in his study comparing "sense of community" in several neighborhoods, found that residents of Greenbelt Homes housing could be characterized as "quite cohesive." He concludes that "beyond the affordability of housing and convenient location to be found in Old Greenbelt, many residents come to like it further and wish to continue living in the neighborhood because of the strong friendships and bonds of attachment that are formed through the process of living there."[22] Many would agree that this feeling exists in much of the center of town, not just in Greenbelt Homes housing.

More than physical proximity is involved in identification with the community. Home ownership forms an important part of that identity. Families who purchase homes make a financial commitment to the community, plan to stay for a while, and are therefore likely to become involved in community activities. Apartment dwellers, frequently transient, with little long-term interest in their physical surroundings or desire to participate in local affairs, seldom contribute to the local community.

Greenbelters recognize these facts. Greenbelt Homes housing must be occupied by owners, both to ensure that they are well maintained and to encourage a sense of community on the part of members. James Giese, as city manager, approved the building of owner-occupied condominiums and town houses in Greenbelt East: "We were pleased because we feel that owner-occupied residences are much more participatory in the community than rental apartments with a high degree of turnover."[23] Having residents who will participate in the community is crucial for Greenbelt, with its strong ethic of cooperation.

Using physical proximity to the center of town and home ownership as important factors, one would deduce that those who own their homes in the center of town identify most strongly with the community. Those in owner-occupied homes in Greenbelt East and in apartments in the center of town would have somewhat less community identity. Residents of Springhill Lake, apartments separated from the center by the Beltway, would feel the least community identity. This analysis is supported by the pattern of voter participation in city elections over the years, in which Old Greenbelters vote most, Springhill Lake residents vote least, with others falling in-between.

For a town whose residents wear T-shirts emblazoned "Greenbelt is Great," the modest physical appearance of central Greenbelt is probably disappointing to at least some visitors. Much of the original housing is still barrackslike in appearance but in many cases is disguised, and improved, by additions and landscaping. The shopping area, Roosevelt Center, always seems to look slightly bedraggled, in great contrast to the sophisticated new shopping areas in Greenbelt East. However, people in the heart of town seem to like the Center the way it is, a familiar, comfortable place. It offers basic stores and services, including several banks, a tailor and a beauty parlor, a High's Dairy store and a variety store, a realtor and Greenbelt Carryout. The owner of Roosevelt Center at one point suggested installing a health spa in the old movie theater, an idea indignantly rejected by the populace in favor of the Cultural Arts Center. The central area, lined with trees and benches and with Lenore Thomas's *Mother and Child* statue at one end, is where the "Center bums" hold sway, sitting for hours talking. The area is most often a resting place for older, retired men, a change from the 1950s, when it was occupied by young mothers, running children, and baby carriages.

The various neighborhoods that compose the center of town have not always been comfortable in their relationships with each other. Just as the

Greenbelt Aquatic Center, 1996 (built adjacent to original
outdoor pool in matching neo art deco style).
Photo courtesy *Greenbelt News Review*

people in the "defense homes" experienced some difficulties in being accepted by those in the original housing, some in Boxwood Village and Lakewood felt as if they were not quite equal to the earlier residents. In 1966 one
new Boxwood Village resident wrote a letter to the editor regarding the
neglect of her area of town not only by the paper but also by other residents.
She concluded, "I like this Greenbelt that has been yours for so long and is
now becoming mine and my neighbors'—but most of all we hope you'll like
us." She realized what would demonstrate their successful integration into
town life: "We may even have a mayor from Boxwood Village at some future
date!"[24] This came true when Gil Weidenfeld, resident of Boxwood Village,
began a lengthy service as mayor in 1975, signaling the acceptance of "newer"
people in positions of authority.

Even residents in the original housing sometimes suffered from a feeling
of being not quite equal if they were not among the "first families." Those
moving into Greenbelt Homes housing in the 1950s sometimes felt slightly
second-rate because they were not among the "pioneers." However, current
residents always welcomed newcomers into town activities, as more help was

always needed and appreciated, thus leading to the blending of newcomers into the Greenbelt family.

At the time of Greenbelt's origin, during the Great Depression, one of the qualities residents most liked about the town was the similarity in income among families. All were equally poor and assisted each other whenever they could. It was considered in bad taste to flaunt possessions that others might not have. When a family moved out of Greenbelt Homes housing to a bigger home and yard in Boxwood Village or on Lakeside Drive, it seemed to demonstrate too much concern with status. Some of this attitude still remains, as revealed by residents' feelings about the building of new, expensive homes on Ridge Road. People desiring such homes are considered "uppity," trying to be better than others. This negative perception of what most Americans would regard as normal upwardly mobile behavior remains strong, especially in the central part of town, where the Greenbelt ethic is most deeply felt.[25]

In a number of ways individual neighborhoods in the center of Greenbelt replicate the actions of the town as a whole. In June 1982 residents of Woodland Hills celebrated the twenty-fifth anniversary of their neighborhood with a picnic. One Woodland Hills resident noted, "A neighborhood is a rare item in today's society. It doesn't occur by itself. It has to be cultivated by those who live there. That is why Woodland Hills must be continued."[26] Boxwood Village residents held a twentieth anniversary party in August 1985 in the form of a gala dinner-dance at the American Legion Hall. Dinner speakers told the history of the area, beginning with the 1960s, when families with young children moved in, and the neighborhood sponsored dances, bowling leagues, and sports events. During the 1970s the neighborhood actively fought developers, in cooperation with the city. In recent years original owners have been moving out, seeking smaller homes or apartments, and families with young children are moving in again, stimulating a new round of activities.

The first residential area built outside of central Greenbelt, Springhill Lake, began the separation of Greenbelt into distinct parts. Residents of Springhill Lake have felt, and are, cut off from the rest of town. The many transient residents, especially the large numbers of University of Maryland students, make for a population not highly involved in city matters. However, some residents do become active in town endeavors: two people from Springhill Lake have been the recipients of Greenbelt's Outstanding Citizen of the Year award.

The plan of Springhill Lake is reminiscent of Old Greenbelt. Its apartments surround a central shopping center, and the design includes an elementary school, playgrounds, a pool, inner walkways, a lake, and a recreation center. Some areas are set aside as green space. For many years a community social director helped plan and organize political clubs, civic associations, and social activities, as well as put out a newsletter. In 1969 a Springhill Lake resident ran for and won a seat on the city council, signaling the beginning of the area's integration into the city. Springhill Lake residents celebrated their tenth anniversary in May 1974 in typical Greenbelt fashion, with a baseball game, music, a square dance, a barbecue, and the public recognition of twenty-one families who had lived there for ten years. In its first ten years, residents pursued interests in politics, the civic association, the Parent-Teacher's Association, and activities organized by the social director. Residents organized block picnics and games. From these activities a strong sense of community emerged. The developer of Springhill Lake, Edward Perkins, has observed that visitors have come to Springhill Lake to study its innovative design. In all this, Springhill Lake seemed very much like Old Greenbelt.

However, the community began to change by 1977. Many activists moved away and were replaced by single parents or families in which both parents worked. Fewer people seemed to have time for or an interest in community activities.[27] In 1979 the social director commented on changes in the population: "In the early days Springhill Lake was the home for 'young marrieds,' but now there are an awful lot of one-parent families, singles, college students, and empty-nesters, parents whose children have grown and moved out on their own." A resident explained another change: "In the early days, Greenbelt didn't care about the apartments. Now, Greenbelt takes us into consideration."[28] Even though residents often show little interest in political activities, they enroll readily in city-sponsored classes such as bridge, dance, and aerobics. In January 1985 the Springhill Lake complex was sold. The new owners began a $12.5 million renovation of the apartments and grounds; established a $4,000 scholarship for an Eleanor Roosevelt High School graduate to attend Prince George's Community College; and hired a community relations director to work with residents as well as with organizations outside of Springhill Lake. The marketing director noted, "We have about 9,000 people in Springhill Lake. We really are a city within a city. We're trying to promote a sense of community."[29]

After the completion of Springhill Lake, developers turned their attention to land east of the Baltimore-Washington Parkway, the only large undeveloped space left in Greenbelt. The entire area, comprising a number of subdivisions, shopping centers, and office parks, is commonly referred to as Greenbelt East. Greenbriar "luxury" condominiums, which were involved in the controversy over the Baltimore-Washington Parkway pedestrian overpass, were built first. Because of the overpass, those residing in close-in areas such as Greenbriar can quickly reach the center of Old Greenbelt.

The residential areas of Greenbelt East, in addition to Greenbriar, are Greenwood Village and Glen Oaks, north of Greenbelt Road, and Windsor Green, Greenbrook, and Hunting Ridge, south of Greenbelt Road. Greenbriar and Hunting Ridge are condominiums, Greenbrook consists of large individual homes, and the rest are town houses. Developments in Greenbelt East were the first typical suburban enclaves in Greenbelt, with behavior to match. Greenbriar residents resisted the parkway overpass because they did not want strangers walking through their parking lots. As often happens in modern suburbia, it took the issue of crime to bring people together: in May 1982 Greenbriar residents began the first crime watch program in the city.[30]

In contrast, some areas in Greenbelt East sponsor events reminiscent of Old Greenbelt. In July 1984 Greenbriar celebrated its tenth anniversary with speeches, entertainment, a tree planting, and refreshments. Greenwood Village residents held a block party soon after the neighborhood came into existence, in June 1986. Residents manned a booth at the Labor Day festival and became involved in the fiftieth anniversary celebration. While people living in Old Greenbelt are not surprised to see a banner across Crescent Road reminding them to vote in an upcoming Greenbelt Homes election, it is more unusual to see a large sign posted at the entrance to Hunting Ridge condominiums reminding residents to get in their proxy votes. The Greenbriar Community Association president William Ayers observed in 1984, "At first we felt we did not belong, but with time we became an integral part of the city. Now we feel needed and wanted, and very definitely have a role to play in the city."[31] In 1985 Greenbelt East people formed the Greenbelt East Association, an organization of individual neighborhood groups working together such as the organizations often created by Greenbelters to enhance the groups' effectiveness and cooperation. One of the association's first acts was to take part in the Festival of Lights, after a resident noted, "The City has a living Christmas tree in the center," and went on to ask, "Why not a living tree in Greenbelt East?" A living tree in Greenbelt East was decorated that year.[32]

The Greenbelt East group enlarged to include new areas as they were built, forming the Greenbelt East Advisory Committee in July 1986 to communicate to the city their positions on matters of common interest. In March 1988 the group worked against a developer's zoning appeal that would have increased the allowable density of buildings on several land parcels. They based their opposition on the already-clogged roads in their area. At the annual meeting for election of officers and discussion of new programs in January 1989, the group decided to establish ties with other Greenbelt homeowner associations. The president commented that the Greenbelt East Advisory Committee "has always tried to present itself as a bridge-building organization to other groups in the Greenbelt community."[33] The formation of the advisory committee followed the typical Greenbelt organizational pattern of creating a committee of group representatives to coordinate objectives and actions. It appears that Greenbelters have been able to extend their activist ideals and ideas to new parts of town.

At the request of the city, the Urban Design Section of the Maryland–National Capital Park and Planning Commission undertook a study on ways to unify Greenbelt East. It was presented to residents at a meeting on March 15, 1988. According to the plan, Greenbelt East would be unified by the use of green space in a "ring and spine" system. The ring would be Hanover Parkway and Mandan Road, landscaped with trees along the median strips and roadsides. Existing green areas, including buffer zones for the Beltway and the Parkway, two city parks, and woods within residential areas, would be connected by internal pathways, forming a spine of green space from north to south. Liora Hawmann of the Urban Design Section explained that "the spine would function as an internal pedestrian system independent of roads, similar to the inner walkways of historic Greenbelt."[34] She pointed out that green space in the undeveloped southern portion would be rapidly disappearing, so the city would need to work quickly to retain open space.

To the east of Greenbelt East, not actually within Greenbelt city limits but with a Greenbelt postal code, the National Aeronautics and Space Administration's Goddard Space Flight Center has had a major impact on the city. The 1,100 acre facility employs 3,700 civil servants and more than 8,000 contract personnel, with an annual operating budget of $1 billion.[35] Goddard's existence draws numerous private companies specializing in related technology to locate in Greenbelt's office parks.

In May 1976 Goddard opened a visitor's center and encouraged Green-

Greenbelt Police Station, 1989 (built in neo art deco style to match original city buildings). Photo by J. Henson, courtesy *Greenbelt News Review*

belters to come and find out about their space-age neighbor. Activities at Goddard often bring worldwide attention to Greenbelt. In August 1975, during the joint Russian-American Apollo-Soyuz mission, a special postmark was created for the Greenbelt post office, and requests for its use were received from around the world. Numerous Goddard employees involve themselves in the life of Greenbelt, partly because many of them live in Greenbelt East. Greenbelt activists, never missing an opportunity to recruit volunteers, have convinced Goddard employees to help with such projects as tutoring schoolchildren, sponsored by Greenbelt Cares Youth Bureau.

Greenbelt's many neighborhoods, from Springhill Lake in the west to Windsor Green in the east, will always have difficulty feeling unified as a city due to the massive roadway canyons dividing them. However, one factor not yet mentioned probably helps residents in the different areas become more familiar with each other. Ever since the late 1950s, when Greenbelt Homes residents formed a co-op to build the first nongovernmental homes in town, people have frequently moved from one section of town to another. The early pattern of Greenbelt Homes members moving to Boxwood Village and Lakewood has continued and has spread to other neighborhoods, caused by residents' changing needs. For example, a single person marries and moves from Springhill Lake to Greenbelt Homes housing. When the couple has a family, they buy a house in Boxwood Village. After the children leave home, they purchase a townhouse in Greenwood Village. Such movement occurs frequently and provides people in each neighborhood with neighbors who are familiar with other areas. What they all have in common is the desire to stay in Greenbelt.

Just as development has greatly modified the appearance of Greenbelt, so a change in the image of the town has occurred. In 1937 all residents lived in near poverty, a requirement for residence. However, the government raised the income limits during World War II, and after the town was sold average incomes rose as Greenbelt became a middle-class community, albeit not a typical one. During the 1960s, when fighting against development, Greenbelters continued to be regarded by outsiders as "kooks and Communists" for not supporting the "American way" and for insisting on being different.[36] However, by the 1980s young, urban professionals discovered Greenbelt, especially Greenbelt East. Faculty and staff at the University of Maryland discovered Greenbelt Homes housing, which was not only close by but also affordable. Unlike many other parts of Prince George's County, Greenbelt's housing is well maintained. The city offers amenities not available elsewhere, such as parks and recreation facilities. The addition of a Hilton hotel and Capitol Cadillac also contributed to the city's new image as a highly desirable location. The former city manager James Giese provided an example that illustrates Greenbelt's change in status: in the 1960s, the movie theater advertised itself as being in College Park, hoping that would improve business. Now, a hotel in nearby New Carrollton calls itself the Greenbelt Sheraton. Developers advertise new housing as being in Greenbelt even when it is located outside the city limits with no link to the town whatsoever.

Greenbelt's political influence in the county has changed as much as its image. From being almost nil in the 1960s, Greenbelt's power gradually increased as Greenbelters gained positions in the county power structure. When the results of massive overdevelopment in parts of the county became apparent during the 1980s, the wisdom of Greenbelters began to be appreciated, and the town now has a great deal of influence in the planning and zoning process. An insistence on planning and the desire for green space are now county policy.

The development of Greenbelt finally ceased due to lack of available land. The division of the town into eight pieces, divided by freeways, made it increasingly difficult for the city to function cohesively. A number of Green-belters, especially in Old Greenbelt, with their background of cooperation and tradition of community activism, refused to accept the division of their town and determined to take action. Working with city planners, they hoped to unify the town physically, based on what already existed.

A crucial physical link came into existence with the advent of Eleanor Roosevelt High School, followed by the Baltimore-Washington Parkway overpass. Residents in Greenbelt East could now walk to the Center in a few minutes. Working with the plan already formulated to unify Greenbelt East with a system of walkways, city planners expanded this into a citywide system of hiker-biker trails, which reaches all areas, crossing the parkway overpass.

The opening of the Green Line of the Metro (the Washington area subway system) on December 11, 1993, provided Greenbelt an easy link with down-town Washington. Greenbelt is the terminus of the Green Line, which was the last to be built in Metro's 103-mile system. However, its effect on Green-belt was minor compared to that of the District of Columbia and other suburbs, where the building of a Metro station spurred major new develop-ment. Both the station and the vacant land around it are just outside Green-belt city limits. The focus of the station, with 3,500 parking spaces and its own access ramp off the Beltway, is to lure drivers off the Beltway and onto the subway.[37] The building of the Greenbelt Metro station led to a further unify-ing of the city. A Greenbelt shuttle bus that stops at the Metro station circu-lates throughout the city in a figure eight. One loop goes to Springhill Lake and the Metro station, the other travels through Greenbelt East. The loops touch at the Center.

Another mechanism for bringing together people from different parts of town, providing an opportunity for them to get to know each other, are the many advisory boards that deal with the city council. The city council initi-ated the first advisory board, the Advisory Planning Board, in 1961 when the council felt the need for specialized city planning advice. The purpose of all of the advisory boards is to advise the city council; they have usually been created in response to specific problems. The city council selects board mem-bers from applicants, who are informed of vacancies via advertisements placed by the city in the *News Review*. The various boards and committees provide an easy way for citizens to become part of the city's political process and to influence actions taken by the council. Residents of Greenbelt East and Springhill Lake utilize them to learn about city politics and to gain influence for their parts of town. Some members of the boards and committees have run for, and served on, the city council.

Attempts to link the city together physically, a determination to keep the city boundary intact, efforts to involve all areas in local politics, and providing the newspaper to each resident so all can get a weekly dose of Greenbelt ideology, these work to create and retain communal identity. Activists are committed to keeping and enhancing community identity so that their cooperative ideals can be maintained. Who exactly are these activists, who do all the work to create community? These individuals function as leaders but often without holding elected or official positions. In fact, they frequently prevail in challenges to city council members, bringing about changes in the behavior of those in office.

Individuals in Greenbelt commonly get drawn into active involvement in town affairs by participating in co-ops, whether the nursery school, the grocery store, or housing for the elderly. A certain number of people who are drawn into one activity go on to become active in other endeavors. The sense of responsibility that many Greenbelters feel also plays a key role in getting people involved. A comment about Elaine Skolnik, at the time of the fiftieth anniversary celebration, provides an example: "She works so hard for Greenbelt that it prompts all of us to ask ourselves how we can serve the community. And, as many of us can attest, when Elaine needs help, for the *News Review* or any other community project, her friends and admirers are there to back her up. How can we do less when she does so much?"[38]

Leta Mach, in her interviews with the winners of the Outstanding Citizen of the Year award for the fiftieth anniversary oral history project, detected certain qualities these individuals had in common. She characterized them as mature, modest, with a quiet sense of self-confidence and a strong feeling that what they were doing, no matter which activity they devoted themselves to, was important to the life of the city.

Observers of community life in Greenbelt express fears about the future supply of citizen activists. In the past, women with children at home formed a large portion of the volunteers. In the 1950s, the editorship of the *News Review* passed from one mother to another, depending on who was not having a baby. Men got heavily involved in politics, sports, and other activities, spending much time after work away from home. Now, with increasing numbers of single parents, or parents who both work, few have free time for community involvement. In recent years an increasing supply of retired individuals has filled the void. However, some worry that with younger people not being drawn into community affairs, there will be no one to replace current activists.

One solution to this problem consists of the Greenbelt network, described to me this way: "Everybody took care of everybody. No matter where you went in the world, our kids wouldn't get into trouble because somebody knew somebody from Greenbelt. You could be on a subway in New York and mention Greenbelt and somebody would say, I know somebody from Greenbelt. There were always people who knew people. It was a network of protection and concern. This is what they're protecting in Greenbelt. It still exists."[39] When people leave Greenbelt, they often maintain ties to those left behind. Common patterns include returning for activities in favorite social organizations or remaining a member of a Greenbelt church. Many come back specifically for the Labor Day festival. Groups of former Greenbelters in Florida hold reunions.

The aspect of the Greenbelt network of significance here is the return of children raised in Greenbelt. After leaving home for school or jobs, many come back to raise their families in the town where they grew up. They carry on town traditions, regarding active participation in community life as normal because that is how they remember family life while growing up. This segment of the community, with its remembrance of things past and its willingness to carry on, holds the most promise for the continuation of Greenbelt as a community of active cooperators.

While busily maintaining traditions and the physical integrity of their original buildings, Greenbelters have witnessed many changes. The population of the city reached 21,096 in 1990, while that of Prince George's County was 729,268 (see table A.1). As vacant land is virtually gone, it is now apparent that Greenbelt will not reach the 50,000 population once projected for the city by planners and developers.

Recent changes in Greenbelt's population reflect both national and Washington metropolitan area demographic trends. Household size continues to shrink, while the number of minorities and the elderly increases. The golden-agers organize themselves into groups such as the Gray Panthers and are quite vocal about their needs. The greater presence of minorities is apparent more in the fringes of town than in Old Greenbelt. Most Hispanics reside in Springhill Lake, while Asians, many of whom work at NASA, live in Greenbelt East. For many years members of minority groups have appeared in positions of prominence in town life. An African American woman won the Miss Greenbelt contest at the Labor Day festival in 1976, a fact accepted as a matter of course and only made apparent by her picture in the *News Review*. The president of Citizens for Greenbelt in 1989–90 was a black man, while a man of Hispanic origin living in Springhill Lake recently received the honor of Outstanding Citizen of the Year.

Greenbelters have always prided themselves on their acceptance of others, and this remains an important part of their self-image. A Boxwood Village reader of the *News Review* wrote in the February 24, 1966, issue of the paper of her delight in finding Norwegian, Italian, Jewish, and Irish neighbors. Greenbelters have been working to accept additional groups ever since. The latest Greenbelt Homes informational brochure prominently displays a picture of several blacks looking at housing. University of Maryland foreign students residing in Springhill Lake provide another element of diversity. Already home to foreign diplomats and World Bank employees, the Washington area is absorbing an influx of poorer and less well-educated immigrants, especially Hispanics and Asians. Increasing numbers will presumably continue to find their way to Greenbelt.

The way Greenbelt residents express their feelings about their community has not changed over the years: they stress small-town neighborliness and personal connections. The following statements were made in the 1970s:

> Even though it's not very small, it has the kind of neighborliness you think of in small towns. [1973]

> It's like a little southern hospitality has moved up north. People are friendlier, willing to stop and talk to you and say "hi." You don't get the cold shoulder the way you do in other cities. You know the people in the stores. Everybody knows everybody that's been here for any length of time. [1976]

> This isn't just a suburb where people go at evening and close the doors. Activities make the community more vital and the people more vital. And we squeeze it to the utmost in Greenbelt. [1978][40]

In the 1980s residents noted their increasing differences from the mainstream politics and government of the time, while continuing to stress the small-town feel of life in Greenbelt. The May 14, 1981, *News Review* carried these comments: "It's a haven from Reagan Washington. Its still a liberal island. Eleanor Roosevelt would still be the idol here." "It's one of the few places that went for McGovern. Being at the polls in Greenbelt on election day is almost like a social occasion." In 1982 the city manager James Giese remarked:

> Sure I've thought of leaving, but I never really felt I would get that much out of a change. There's a good respect for the government here. People who work for the government feel appreciated. There's more friendliness, less putting on airs here. Greenbelt has many characteristics of a small town. . . . There's a community spirit that dates back to the beginning, I think. People here were ostracized. People said "They're all poor people" or "They're all Communists." So the people living here stuck together.[41]

Community newcomers during this period could clearly explain their feelings:

> We had a lot of options before we moved here, but we liked this community. We liked the idea of being able to walk to a lot of places, we liked the price here, and we liked the idea of living in a co-op. There's something slightly unconventional about it. We didn't have the money or the interest to move into the typical suburban environment.

> You have a real neighborhood here. You can walk to the tennis courts, walk to the library. . . . I don't ever want to leave.[42]

Those attending the forty-fifth anniversary celebration in 1982 remembered Greenbelt over the period of its existence:[43]

> [Greenbelt was] real small-town America; the most beautiful example of the ideals on which this country was founded.

> Greenbelt residents, then and now, practice democracy with a vengeance. Someone once asked us what we do out in this garden paradise. The answer: have meetings.

> Greenbelt is still a strangely heterogeneous community. It is the antithesis of little boxes. There is total acceptance by the sheet worker for the Ph.D. historian next door, or the town clerk for the gas station attendant. Keeping up with the Joneses is just not a phenomenon in Greenbelt.

> Class lines were blurred. There was an incredible closeness, fostered by community activism and a lot of block parties. Greenbelt was an experience we should look at today.

> We held very traditional American values: high on democracy and hard on communism. Greenbelt was radical only in that it was a "unique humanistic endeavor" to improve the social welfare of everyone in the town. There was a sense of community. The friendships we made lasted all these years.

These comments focused on Greenbelt's traditional elements, emphasizing the town role as a showplace for the premier American value, democracy.

Southern Maryland U.S. District Courthouse (adjacent to
Capital Office Park, across Kenilworth Avenue from
Greenbelt Center). Artist's rendering courtesy
*Greenbelt News Review*

This occurred in a town regarded as "modern" and experimental. This di-
chotomy, looking forward and backward at the same time, did indeed exist.

Comments made at the time of the fiftieth anniversary celebration in 1987
reflected on the obvious physical changes, while observing that the spirit of
the community remained the same. This was typical: "Greenbelt has changed
physically. It's more urban; we now have a skyline, but the people and the
attitude and feeling about the community has not changed. It's still a caring
union and there's a desire to be active in whatever issues come up."[44]

Comments by those who have moved to the town within the last ten years
show that they have absorbed both Greenbelt's ethic and its history:

> When I went to the Labor Day Festival I thought I had been transferred back to a
> small town in Iowa. It looked exactly like the carnival they had every spring, the
> floats, the cotton candy, carnival booths. It could have been lifted out of small-
> town Iowa, but, if you listened carefully, you could hear the Beltway in the back-
> ground. . . . The sort of people who were willing to take the giant step to go to a new
> planned community were likely to be the kind of people who were at least tolerant
> of new ideas, and I think you still see that. . . . The sense of community will
> continue. Greenbelters are a feisty bunch and will continue to be different from
> other people—at least I hope so.[45]

> When I was looking for a place to live in PG County, people told me about
> Greenbelt. They described it as a place where people are born, raise their children,
> and die, where people put down roots. And here I am.[46]

Feelings about life in Greenbelt have thus remained remarkably stable over time, as newcomers describe Greenbelt in terms almost identical to those used by the pioneers. They assimilate the philosophy of cooperation and accept the desirability of being active in their town. Like the original tenants, new residents appreciate the town layout, which encourages a neighborly walking community.

The quotations above are evidence that Greenbelters feel a fondness for their town on an emotional level, an appreciation that goes beyond such tangible factors as structures and institutions. The town's central core remains almost untouched by time, which is why people remark that the town is like "something out of the fifties." However, the ring around the core, now composed of office parks, condominiums, town houses, shopping centers, and freeways, is a distinct contrast to the worn-out tobacco fields that were there in the 1930s.

This ring, which was originally to remain a permanent greenbelt to limit growth and overcrowding, was developed when Greenbelt Homes sold the vacant land, being unable to pay the taxes on it. Developers eagerly purchased pieces of the greenbelt to pursue their own plans. Prince George's County government controlled land zoning and generally acceded to the wishes of developers. It took every effort of Greenbelters to maintain the central core in a relatively undisturbed manner. In recent years, the city has purchased the remaining undeveloped parcels of land in an effort to restore as much of the greenbelt as possible. For parts of the area, this effort was too little, too late. Thus Greenbelters had mixed success in the physical maintenance of their planned community, maintaining the original buildings but not the greenbelt. They did retain the ideology of a planned community as their guiding principle. The neighborhoods surrounding Old Greenbelt frequently organize activities similar to those of the original part of town, so even though Greenbelters lost their greenbelt, they successfully carried their ideology and principles to the surrounding area. Expanding even farther, the influence of the city reached officials of Prince George's County, who now use Greenbelt's planning principles as their guidelines.

Over its history, Greenbelt's cooperative ideology flourished and spread. Economic cooperation went full circle, as Greenbelt Consumer Services expanded far beyond the town and Greenbelters lost all influence over it. After expanding too rapidly, the GCS ran into financial difficulties and was sold. Greenbelt residents responded by reestablishing the co-op in its original form. Other financial cooperatives such as Twin Pines Savings and Loan (taken over by Mellon Bank after the savings and loan crisis) and the credit union continue to prosper. In addition, town residents applied the cooperative concept to Greenbelt Homes and other housing groups, nursery school, day care, health care, kindergarten, senior citizen housing, and the newspaper. Residents also used the cooperative concept politically, quickly learning that their united influence made them more likely to achieve their goals. Greenbelters maintained the ideal of cooperation in spite of the surrounding physical change.

The town newspaper and the city government both played crucial roles in the maintenance of Greenbelt and its principles. Having their own city government, Greenbelters could control their destiny in matters under their jurisdiction, which then reinforced their will to do so. The *Cooperator* and then the *News Review* provided not only information but also a way for Greenbelters to communicate with each other. Citizen activists could use the newspaper to involve large numbers of residents in an issue whenever it seemed necessary for the good of the community.

Physical boundaries play an important part in delineating a community and have done so in Greenbelt, where physical barriers have worked both for and against the maintenance of community identity. Clarence Perry's neighborhood unit, an important part of Old Greenbelt's design, proved to be cohesive, as predicted, but made it somewhat difficult for newer sections to merge with the old. The major roadways surrounding the central part of town form a boundary and have helped to maintain the identity of Old Greenbelt as a distinct unit, but this boundary made it extremely difficult for Greenbelt East and Springhill Lake to be part of town. That the disparate parts of Greenbelt have been brought together as much as they have is due to the efforts of the *News Review* to reach all residents, physical links such as hiker-biker trails and buses, and the fact that these communities are united under one, very active, city government.

To understand how the Greenbelt principles—a belief in a planned, cooperative community—have been maintained, it is necessary to start with the original settlers. They were well educated and upwardly mobile. They were active in their search for a better life and open to new ideas, as evidenced by their application for residency in a town that was part of a federal experiment and by their willingness to follow the procedures necessary to become residents. They agreed to live in and be part of a planned community, to work with and for the group. Thus the town began with a much larger proportion of people willing and able to be active in community affairs than would be true of most towns.

As the years passed, the above characteristics led logically to the formation of others. As residents worked together they gained knowledge of how to organize effectively to achieve their desired results. When they realized they could organize effectively and beheld the positive outcomes, they developed a desire for self-determination in community affairs. In the crucial early years, federal government officials and the staff of the *Cooperator* actively encouraged this process. A Greenbelt philosophy of life developed. Such a philosophy was an implicit goal of the federal program and an explicit goal of the newspaper staff. The Greenbelt philosophy emphasizes the formation of a community identity, which fosters a sense of responsibility to others in the community. Greenbelt's unique beginning has been utilized at every opportunity to reaffirm group identity. A willingness to take action to preserve the group and its beliefs is a key feature. The self-image of residents reveals what they hoped to be: tolerant, accepting, forward thinking, modern.

Greenbelters, in fact, have a number of "firsts" in their history, all sources of pride, starting with the innovative methods and materials used in the original construction. The federal government gave them the first town manager form of government in Maryland, the first recreation department in the state, the first "modern" telephone system in the metropolitan area, and the first kindergarten, nursery school, and public pool in Prince George's County. Greenbelters claimed one of the first public libraries, Boys Club, and school lunch programs in the area. Betty Harrington became the first elected woman mayor in the state.

The Greenbelt philosophy is dedicated to democracy in action, wherein each individual is encouraged to participate in local affairs. Thus ironically in a town in which the importance of cooperation is stressed, opportunities abound for individual growth and development, as everyone is urged to contribute whatever and however they can. Books, libraries, education, and the arts are regarded as highly as parks and woods, all being part of a desirable lifestyle. In addition to unique opportunities, Greenbelters have two responsibilities: to stand up for what is morally right and then to take appropriate action; and to convince others of the desirability of the Greenbelt philosophy—that is, the goodness of life in a planned, cooperative community. They have never been shy of attempting the latter, as they are truly convinced of the value of the Greenbelt way of life.

The Greenbelt philosophy began as an artificial construct, given to a group of people, strangers to each other, selected to form a new community. It operated as a self-fulfilling prophecy, providing residents with a goal and a standard of ideals and behavior, which they attempted to reach. They had a sense of what they wanted to be and worked to become it. Greenbelters used their principles to guide their actions, which then became precedents for the actions of others. Greenbelt's reputation as an activist, liberal community draws to it those of a similar bent. Like any suburb, it has attracted people similar to those already there. This influx of like-minded residents helps provide the people necessary to maintain the original goals and ideals.

The Greenbelt principles focus on a life of action and involvement. To have ideals is only the first step; the necessary second step, and much more difficult, is to live in accordance with them. For Greenbelters, this translates into group organization. The basic form of organization is that most democratic of institutions, the committee. Residents realize and admit that committees are sometimes unwieldy and inefficient, but the equality and opportunity for all that they provide makes them necessary. Greenbelters use committees in several ways. The committee of committees is used to bring together individuals representing many organizations to work on a large project. Examples are the Defense Council in World War II, the Greenbelt Recreation Coordinating Committee, formed in 1957, and the steering committees for the twenty-fifth and fiftieth anniversary celebrations.

Greenbelters also form committees to deal with specific issues, examples being the Citizens Charter Referendum Committee of 1958, the Citizens for a Planned Greenbelt, organized in 1965, the Freedom of the Press Committee and the Save Our Community Committee, both created in 1966, the Committee to Preserve Greenbelt's Co-op, 1984, Friends of the *News Review,* 1985, Citizens for North End Center, 1986, and the Save the Greenbelt Committee of 1989. Such a committee generally begins when several concerned citizens place an announcement of its formation in the *News Review,* calling all who are interested to attend the first meeting. The activities of the group depend, of course, upon its purpose; however, certain types of action repeatedly appear. For political protest, Greenbelters organize letter writing, door-to-door petition signing, and mass meetings. Door-to-door campaigns have also proved useful in fund-raising.

When Greenbelters divide on an issue, which occurs frequently, the process is somewhat different. Occasionally committees form representing both

Green Line train, Greenbelt Metro station, 1993 (greater
Washington, D.C., subway system began Metro Green Line
service to Greenbelt on December 11, 1993). Photo by
J. Henson, courtesy *Greenbelt News Review*

sides of the issue, such as occurred in the debate about the future use of
Center School. A stream of letters to the editor appears, testimony flows in
city council meetings, public meetings give everyone a chance to have their
say, flyers materialize on doorsteps, advertisements appear in the *News Review,* and a vote is held to determine the outcome.

Developing a new co-op is easy in Greenbelt, because so many know how
to do it. Similarly, the process necessary to form an organization and to
achieve its goal is common knowledge. Greenbelt teenagers, who have appeared in large groups at city council meetings to demand their rights, demonstrate that such knowledge is learned behavior. That many of these young
people remained in, or returned to, Greenbelt in adulthood ensures that the
next generation will observe their organizational skills and benefit from their
example.

One form of political activity, the act of going to the polls, deserves some
comment. Greenbelters regard voting in city and national elections as both a
sacred right and a public responsibility, and high percentages of residents
vote. Further, if a particular issue stimulates interest in a local election,

whether for the city council or the Greenbelt Homes board of directors, higher than average numbers of people place themselves on the ballot and many more than usual turn out to vote. Many Greenbelters feel strongly about issues and dispute fiercely with one another on an issue, until they decide that one and move on to the next. In spite of, or perhaps because of, their intense feelings, which provoke disagreement, they maintain their strong communal identity. If challenged from the outside, they unite quickly and efficiently to uphold their community.

The characteristic behavior of Greenbelters emanates directly from the ideals of community espoused by federal officials during the New Deal. The ideas used by federal workers to create Greenbelt were not new, just newly collected by the Resettlement Administration to form the greenbelt town program. Another federal government agency, the Federal Housing Administration, created in 1937, is responsible for spreading some of these ideas, particularly the idea of the neighborhood unit, to developers of postwar suburbia. The FHA produced booklets for builders advising them on technical matters, land usage, and architecture; some of the suggestions were straight from the green town program.

In 1938 the FHA published Technical Bulletin 7, *Planning Profitable Neighborhoods*. The bulletin suggested that major traffic arterials surround the neighborhood, while streets within the neighborhood be used for local traffic only. It also suggested gently curving streets. Featured were centrally located commercial, school, and church sites, with the inclusion of parks and the preservation of natural features whenever possible. Similar ideas appeared in Technical Bulletin 5, also in 1938, titled *Planning Neighborhoods for Small Houses*. The FHA began another series with the same suggestions in 1941 with its Land Planning Bulletin 1, *Successful Subdivisions*.[47]

In subdivisions during the 1950s the idea of the neighborhood remained conspicuous enough that builders included a community facility, often a school, as a central focus.[48] Later, the idea of neighborhood and communal facilities seemed to disappear, leaving the suburbia of today with its ubiquitous curving streets that go nowhere. In the 1990s, however, similar ideas have reappeared, as more and more people express their dismay with suburban living as usually formulated. Philip Langdon, in *A Better Place to Live: Reshaping the American Suburb*, analyzes the errors in current suburbs and considers what the solutions should be:

> Suburbs, I believe, need to break free of the crippling limitations imposed on them since approximately the 1940s. We need to develop suburbs that foster neighborhood and public life rather than squelch it. This requires designing suburbs so that public areas . . . are enjoyable to occupy. Comprehensive networks of sidewalks are essential. Sidewalks and streets should be organized so that people will have incentives to explore their neighborhoods. Houses need to treat the public areas as important, congenial places. Instead of glorifying interior and private spaces while leaving the public environment dominated by uncommunicative façades and garage doors, we need to reorient houses so that they dignify and enliven the places where neighbors and strangers come in contact with one another.
>
> We need to rethink our planning ideas so that neighborhood stores, neighborhood institutions, neighborhood gathering places will have a better chance of coming into being and giving heart to the community. . . . We ought to ponder the advantages of creating enough concentration to nurture a vigorous community spirit and to support stores and institutions within walking distance of homes.[49]

Planners working in this new area of neotraditional development focus on TND, traditional neighborhood development. Begun largely by the husband-wife team of Andres Duany and Elizabeth Plater-Zyberk, the idea of TND is spreading rapidly, not just among planners but also among developers. Peter Katz, in *The New Urbanism: Toward an Architecture of Community,* describes a number of projects now being built, both suburban and urban, using TND principles. He enumerates the principles of an ideal neighborhood design:

> 1) The neighborhood has a center and an edge; 2) The optimal size of a neighborhood is a quarter mile from center to edge; 3) The neighborhood has a balanced mix of activities—dwelling, shopping, working, schooling, worshipping, and recreating; 4) The neighborhood structures building and traffic on a fine network of interconnecting streets; 5) The neighborhood gives priority to public space and to the appropriate location of civic buildings.[50]

The design features of TND are, of course, the design features of Greenbelt.

It is too soon to know if the latest use of these ideas will continue to grow and spread. One factor in their favor is the sponsorship of these plans. With Greenbelt, the federal government created the program, earning the instant and undying animosity of the private sector. In current TND projects, private developers, encouraged by planners, are leading the way.

Some critics deride these projects as idealistic utopias or as facile façades for the usual suburban, white, affluent enclave. William Fulton, in *The New Urbanism: Hope or Hype for American Communities?* points out that a program that addresses the physical arrangement of neighborhoods cannot realistically be expected to solve all urban and suburban problems.[51] It is clear that the new urbanism and TND do not address the difficult problems of crumbling inner cities and suburban flight. A regional approach that confronts current economic inequalities is necessary to produce substantive change.

In traditional neighborhood development, the planners hope to create a true feeling of community. The planners of Greenbelt intended the same thing, but seen in comparison, TND clearly lacks the idea of cooperation, which was central to Greenbelt's success. In Greenbelt physical form and function mesh in a unique way, with the form of the planned community supporting and enhancing the cooperative lifestyle. New Deal planners created a symbiosis in which the physical design and cooperation mutually sustain each other, a symbiosis that has proven to be long lasting.

This work on Greenbelt ends where it began, with Lewis Mumford's *The Culture of Cities*.[52] Mumford envisioned a town in which planning clearly makes a difference, where foresight creates a place truly enjoyed by its residents. It could be argued that Greenbelt is Mumford's ideal made real.

# Appendix

TABLE A.1
The Population, Greenbelt and Prince George's County, 1940–1990

| Census year | Greenbelt | Prince George's county | Greenbelt population as % of county population |
|---|---|---|---|
| 1940 | 2,831 | 89,490 | 3 |
| 1950 | 7,074 | 194,182 | 4 |
| 1960 | 7,479 | 357,395 | 2 |
| 1970 | 18,199 | 661,719 | 3 |
| 1980 | 17,332 | 665,071 | 3 |
| 1990 | 21,096 | 729,268 | 3 |

SOURCE: U.S. Bureau of the Census.

TABLE A.2
Minority Groups, Greenbelt 1940–1990

| Census year and minority population | Minority population as % of total population |
|---|---|
| 1940 | |
| 14 Blacks[a] | 0.4900 |
| 1950 | |
| 1 Black | 0.0001 |
| 1960 | |
| 0 Blacks | |
| 17 Other | 0.2300 |
| 1970 | |
| 230 Blacks | 1.2600 |
| 146 Other | 0.8000 |
| 1980 | |
| 1,721 Blacks | 9.9000 |
| 431 Asian | 2.5000 |
| 383 Hispanic | 2.2000 |
| 602 Other | 3.5000 |
| 1990 | |
| 4,085 Blacks | 19.4000 |
| 1,901 Asian | 9.0000 |
| 807 Hispanic | 3.8000 |
| 40 American Indian | 0.1900 |
| 266 Other | 1.3000 |

SOURCE: U.S. Bureau of the Census.
[a]Did not live in government housing but in farmhouses built before the federal government purchased the land.

TABLE A.3
The Aging of Greenbelt Residents, 1940–1990

| Census year | Number sixty-five and older | Total population | Sixty-five and older population as % of total population |
|---|---|---|---|
| 1940 | 29 | 2,831 | 1.02 |
| 1950 | 86 | 7,074 | 1.21 |
| 1960 | 202 | 7,479 | 2.70 |
| 1970 | 596 | 18,199 | 3.27 |
| 1980 | 1,014 | 17,332 | 5.85 |
| 1990 | 1,378 | 21,096 | 6.53 |

SOURCE: U.S. Bureau of the Census.

TABLE A.4
Housing Units, Greenbelt, 1940–1990

| Census year | Number of units | % Rental | % Owner-occupied |
|---|---|---|---|
| 1940 | 900 | 100 | 0 |
| 1950 | 900 | 100 | 0 |
| 1960 | a | | |
| 1970 | 6,519 | 68 | 32 |
| 1980 | 8,005 | 60 | 40 |
| 1990 | 9,938 | 56 | 44 |

SOURCE: 1940 and 1950 data from Mary Lou Williamson, *Greenbelt: History of a New Town, 1937–1987* (Norfolk, Va.: Donning Company, 1987); 1970–1990 data from U.S. Bureau of the Census.

[a] Data available by census tract only, not by individual towns.

# Notes

## Introduction

1. John W. Reps, "Public Enterprise and New Towns: An American Tradition Revisited," in *The Contemporary New Communities Movement in the United States*, edited by Gideon Golany and Daniel Walden (Urbana: University of Illinois Press, 1974), 23.

2. Ibid., 24. See also Paul Conkin, *Tomorrow a New World: The New Deal Community Program* (Ithaca: Cornell University Press, 1959).

3. Arnold R. Alanen and Joseph A. Eden, *Main Street Ready-Made: The New Deal Community of Greendale, Wisconsin* (Madison: State Historical Society of Wisconsin, 1987); Zane L. Miller, *Suburb: Neighborhood and Community in Forest Park, Ohio, 1935–1976* (Knoxville: University of Tennessee Press, 1981); Joseph L. Arnold, *The New Deal in the Suburbs: A History of the Greenbelt Town Program, 1935–1954* (Columbus: Ohio State University Press, 1971); Albert Mayer, *Greenbelt Towns Revisited* (Washington, D.C.: Department of Housing and Urban Development, 1968); Susan L. Klaus, *Links in the Chain: Greenbelt, Maryland, and the New Town Movement in America* (Washington, D.C.: George Washington University Press, 1987). More general works on Greenbelt are those of Otis Kline Fulmer, *Greenbelt* (Washington, D.C.: American Council on Public Affairs, 1942); George A. Warner, *Greenbelt, the Cooperative Community: An Experience in Democratic Living* (New York: Exposition Press, 1954); Mary Lou Williamson, ed., *Greenbelt: History of a New Town, 1937–1987* (Norfolk, Va.: Donning, 1987).

4. Lewis Mumford, *The Culture of Cities* (Westport, Conn.: Greenwood, 1970 [1938]), 452.

5. Henwar Rodakiewicz wrote the script; Morris Cannovsky provided narration; Ralph Steiner and Willard Van Dyke directed; Aaron Copland contributed the musical score. Lewis Jacobs, *The Documentary Tradition* (New York: Norton, 1979).

## 1 / Building a Planned Community

1. Michael V. Namorato, *Rexford G. Tugwell: A Biography* (New York: Praeger, 1988); Bernard Sternsher, *Rexford Tugwell and the New Deal* (New Brunswick: Rutgers University Press, 1964).

2. President Roosevelt created the Resettlement Administration by Executive Order 7027 on April 30, 1935. It was included in the Emergency Relief Appropriation Act of 1935. Joseph Arnold, *The New Deal in the Suburbs: A History of the Greenbelt Town Program, 1935–1954* (Columbus: Ohio State University Press, 1971), chap. 2.

3. Ibid., 24–26.

4. Ibid., 42–43; Robert Goldston, *The Great Depression: The United States in the Thirties* (Greenwich, Conn.: Fawcett, 1968), 229.

5. Arnold, *The New Deal in the Suburbs,* 42–43.

6. Diary Notes 1935, Box 32, Tugwell Collection, Franklin Roosevelt Presidential Library, Hyde Park, N.Y.

7. Clarence S. Stein, *Toward New Towns for America,* 3d ed. (Cambridge: MIT Press, 1966), 127.

8. Clarence Arthur Perry, "The Neighborhood Unit," in *Neighborhood and Community Planning* (New York: Committee on Regional Plan of New York and Environs, 1929), 7:34. Perry's work with the Regional Planning Association of America is mentioned in Roy Lubove, *Community Planning in the 1920s: The Contribution of the Regional Planning Association of America* (Pittsburgh: University of Pittsburgh Press, 1963).

9. Stein, *Toward New Towns for America,* 150. National Archives II, College Park, Maryland, has in its holdings much material on Greenbelt's design, including Hale Walker's neighborhood units.

10. Lubove, *Community Planning in the 1920s,* 47.

11. K. C. Parsons, "Clarence Stein and the Greenbelt Towns," *APA Journal* (Spring 1990): 165–66.

12. Ibid., 167–68.

13. Paul Conkin, *Tomorrow a New World: The New Deal Community Program* (Ithaca: Cornell University Press, 1959), 69.

14. *Looking Back: Greenbelt Is 50, 1937–1987* (Greenbelt, Md.: City of Greenbelt, 1987).

15. *Cooperator,* April 4, 1940.

16. *Cooperator,* May 18, 1938.

17. *Cooperator,* April 13, 1939.

18. Conkin, *Tomorrow a New World,* 202.

19. Joseph G. Knapp, *The Advance of American Cooperative Enterprise, 1920–1945* (Danville, Ill.: Interstate Printers and Publishers, 1973), 225–26.

20. Tugwell #2, Box 668, Eleanor Roosevelt Papers, Franklin Roosevelt Presidential Library, Hyde Park, N.Y.

21. Rexford Tugwell, 1934–1936, Box 73, Presidential Secretaries File, Franklin Roosevelt Presidential Library, Hyde Park, N.Y. Although four communities were originally planned, only three were built. Greenbrook, New Jersey, was halted by court suits brought by local residents opposed to the project.

22. Folder 1, Box 1, Lansill Ms. Collection, University of Kentucky Library, Lexington; *Architectural Record* (September 1936): 219, lists Douglas D. Ellington as architect in addition to R. J. Wadsworth.

23. David Myhra, "Rexford Guy Tugwell: Initiator of America's Greenbelt New Towns, 1935 to 1936," *AIP Journal* (May 1974): 184.

24. For more details on the construction of the town, see Mary Lou Williamson, ed., *Greenbelt: History of a New Town, 1937–1987* (Norfolk, Va.: Donning, 1987), chap. 2.

25. In 1938 a private developer leased land for the construction of a development, comprising ten one-family houses, called Parkbelt. He subsequently sold his interest to private individuals. See ibid., 34.

26. Ibid., 33.

27. Ibid., 34.

28. Ibid., 36.

29. Arnold, *The New Deal in the Suburbs*, 95.

30. Ibid., 36.

31. Ibid., 55.

32. Public Relations, Greenbelt Project Final Report, 2, Folder 5, Box 1, Lansill Ms. Collection.

33. Arnold, *The New Deal in the Suburbs*, 110–11. For details on the cost of the "relief" part of Greenbelt's construction, see 117–19.

34. "Who Is Forgotten?" *Washington Times,* October 18, 1935.

35. "Tugwell's Folly," *Washington Post,* October 13, 1935; Greenbelt Project Final Report, 1, Lansill Ms. Collection.

36. Director, Prince George's County Historical Society, interview by author, April 28, 1990.

37. Ibid.; George H. Callcott, *Maryland and America: 1940–1980* (Baltimore: Johns Hopkins University Press, 1985), chap. 1.

38. Sternsher, *Rexford G. Tugwell and the New Deal*, 228.

39. Two scrapbooks of articles criticizing Tugwell, along with his replies, are in Box 117 and Box 118, Tugwell Collection.

40. "Tugwell's Folly," *Washington Post,* Oct. 13, 1935.

41. Henry S. Churchill, "America's Town Planning Begins," *New Republic,* June 3, 1936, 97.

42. The *Literary Digest* published two glowing reviews of Greenbelt: Frances Fink, "Cooperative Corners," October 16, 1937, and Frances Fink, "Moving Day," November 6, 1937; "American Housing: A Failure, a Problem, a Potential Boon and Boom," *Life,* November 15, 1937, 45; Felix Belair Jr., "Greenbelt: An Experimental Town Starts Off," *New York Times Magazine,* October 10, 1937, 3.

43. "Government Backs Cooperatives," *Business Week,* September 11, 1937; "Co-Op Transfer, Government Housing Project in Maryland Will Take over Filene's Stores, Now in the Black," *Business Week,* July 29, 1939; "Greenbelt Takes Over," *Business Week,* February 3, 1940; Cedric Larson, "Greenbelt, Maryland: A Federally Planned Community," *National Municipal Review* (August 1938): 413–20; "Greenbelt's Problems: Dogs and Transportation," *Christian Science Monitor,* February 11, 1938, 4; Gordon Eames Brown, *Washington Star,* September 15, 1936; "Greenbelt Towns," *Architectural Record* (September 1936): 215–22; John Drier, "Greenbelt Planning," *Pencil Points* (August 1936): 2–13.

44. Sternsher, *Rexford G. Tugwell and the New Deal*, 292.

45. Rexford Tugwell, "The Meaning of the Greenbelt Towns," *New Republic,* February 17, 1937; Rexford Tugwell, "Magical Greenbelt Is Rising, Model Maryland Community," *Work,* District of Columbia WPA (October 1936). See also Theodore C. Wallen, "Tugwell Defends New Deal in Plea at Editors' Dinner," *New York Herald Tribune,* April 22, 1934; "Tugwell Protests Published Figures," *New York Times,* November 20, 1935.

46. Wallace Richards to Mrs. Franklin Roosevelt, October 28, 1937, Box 705, Eleanor Roosevelt Papers.

47. Alexander, Will, 1937, Folder; and Memo from FDR to ER, November 6, 1937, both in Box 705, Eleanor Roosevelt Papers.

48. Will Alexander was assistant administrator to Tugwell in the Resettlement Administration during 1935–36 and thus was familiar with Greenbelt and Tugwell's goals. Alexander had been a Methodist minister and had also worked in the Roosevelt administration in 1934–35 as a member of the Commission of Minority Groups in Economic Recovery. With such a background, he would presumably be well disposed toward the cooperative experiment of Greenbelt. Albert Nelson Marquis, ed., *Who's Who in America,* vol. 21 (Chicago: A. N. Marquis, 1940).

49. *Cooperator,* December 15, 1937.

50. Greenbelt Housing Project Folder (USDA press release), Box 112, Morganthau Papers, Franklin Roosevelt Presidential Library; Harry S. Sherwood, "RA Utopia Chiefs Expect to Turn 'Em Away When 'For Rent' Signs Go Up," June 17, 1937 (article in unnamed newspaper), Folder 1, Box 4, Lansill Ms. Collection; Arnold, *The New Deal in the Suburbs,* 137.

51. Arnold, *The New Deal in the Suburbs,* 142–43.

52. William H. Form, "The Sociology of a White Collar Suburb: Greenbelt, Maryland" (Ph.D. diss., University of Maryland, 1944), 58. Information retained by the federal government gives little indication of the selection process for tenants. Form interviewed Farm Security Administration officials; his information is used here.

53. Joe and Peg Loftus, oral history, Oral History Subcommittee, Greenbelt Fiftieth Anniversary Committee; videotape in private collection.

54. Form describes the liberal and conservative groups but was puzzled by their existence, not realizing their division into two age groups. This was made clear to the author only by discussion with original residents of Greenbelt who participated in the early political battles.

55. Arnold, *The New Deal in the Suburbs,* chaps. 2, 6.

56. Ibid., chap. 9.

57. Greenbelt Miscellaneous, Memo from Roy S. Braden, community manager, to "Fellow Greenbeltians," Folder 1, Box 4, Lansill Ms. Collection.

58. Williamson, *Greenbelt,* 73.

59. Louise Steinle Winker and Hulda Bomburger, in *Looking Back,* 77, 15.

## 2 / Creating a Cooperative Community

1. Frances Fink, "Moving Day," *Literary Digest,* November 6, 1937, 15.

2. *Looking Back: Greenbelt Is 50, 1937–87* (Greenbelt, Md.: City of Greenbelt, 1987).

3. Joseph Arnold, *The New Deal in the Suburbs: A History of the Greenbelt Town Program, 1935–1954* (Columbus: Ohio State University Press, 1971), 167.

4. Peter J. Carroll, "How Greenbelt Co-ops Originated," *Greenbelt Cooperator,* Charter Day edition, June 1, 1938, 24.

5. Bruce Bowman, in *Looking Back.*

6. Mary Lou Williamson, ed., *Greenbelt: History of a New Town, 1937–1987* (Norfolk, Va.: Donning, 1987), 76–78.

7. Williamson, *Greenbelt,* 75–78.

8. To follow the progress of the development of the *Cooperator,* see weekly issues in Tugwell Room, Greenbelt Public Library.

9. Williamson, *Greenbelt,* 77.

10. The floor plans of the interdenominational church are in the Tugwell Room.

11. Williamson, *Greenbelt,* 75.

12. The weekly *Cooperator* provided detailed information on each religious group, including where and when they met.

13. Original teachers at Center School, oral histories, Oral History Subcommittee, Greenbelt Fiftieth Anniversary Committee; videotape in private collection.

14. Two original Center School students, interview by author, June 5, 1989.

15. Students of Center School who then attended Greenbelt High School, interviews by author; teachers who taught the first several years at Center School and Greenbelt High School, oral histories, Oral History Subcommittee, Greenbelt Fiftieth Anniversary Committee; videotape in private collection.

16. Descriptions of special events from weekly issues of the *Cooperator.*

17. William H. Form, "The Sociology of a White Collar Suburb: Greenbelt, Maryland" (Ph.D. diss., University of Maryland, 1944), 50.

18. Although elections were planned for every two years, they were held in both 1937 and 1938 because the town administration felt that newcomers, who were still moving in, deserved a chance to help choose the city council and mayor. It is not possible to compare voting patterns in Greenbelt to those in other towns; statistics available for Greenbelt provide the percentage of registered voters who voted, while those for the United States, such as provided by the Bureau of the Census, give the percentage of those of voting age who voted. Data on percentage of both total population and registered voters who voted is available only for 1970 to 1990 and only for some states.

19. Form, "The Sociology of a White Collar Suburb," 13.

20. Ibid., 433.

21. Hull informed the town manager Roy Braden of his ruling by letter, reprinted in the *Cooperator,* October 17, 1940.

22. Weekly issues of the *Cooperator* noted visitors "of distinction."

23. "Greenbelt's Problems: Dogs and Transportation," *Christian Science Monitor,* February 11, 1938, 4; Christine Sadler, "Greenbelt, One Year Old, Pays Social Dividends as Test-Tube City," *Washington Post,* October 9, 1938, 6; "Greenbelt Bans Display of Wash After 4 PM and on Sundays," *Washington Star,* March 30, 1938; Greenbelt Miscellaneous, Memo from Roy S. Braden, community manager, to "Fellow Greenbeltians," Folder 1, Box 4, Lansill Ms. Collection, University of Kentucky Library, Lexington.

24. Frank Luther Mott, *American Journalism: A History, 1690–1960* (New York: Macmillan, 1962), chap. 27; James T. Patterson, *Congressional Conservatism and the New Deal* (Lexington: University of Kentucky Press, 1967), chap. 10; Katie Louchheim, *The Making of the New Deal* (Cambridge: Harvard University Press, 1983), 12–13.

25. Oral histories, 1981–85, students in oral history classes at the University of Maryland; oral histories, 1982–87, by Dorothy Lauber, executive secretary to the city manager

(published in *Looking Back*). Oral histories, Oral History Subcommittee, Greenbelt Fiftieth Anniversary Committee, videotapes in private collections. A number of interviews of original residents who remain in the Washington, D.C., area were by author.

26. Ben Rosenzweig, in *Looking Back*, 57.

27. Original resident, interview by author, July 17, 1989.

28. Original resident, interview by author, July 17, 1989; Pauline Trattler, in *Looking Back*, 88.

29. Alexa Lauber Kiefner, Henry L. Trattler, Michael Burchick, and Ben Goldfaden, in *Looking Back*, 46, 87, 65, 28.

30. Betsy Wood Walters, in *Looking Back*, 96.

31. Early resident who still resides in Greenbelt, interview by author, May 26, 1989.

## 3 / The War Years

1. Material on characteristics of resident families, 1940–42, from William H. Form, "The Sociology of a White Collar Suburb: Greenbelt, Maryland" (Ph.D. diss., University of Maryland, 1944).

2. Ibid., 143, 144.

3. Ibid., 394–401.

4. Joseph Arnold, *The New Deal in the Suburbs: A History of the Greenbelt Town Program, 1935–1954* (Columbus: Ohio State University Press, 1971), 229.

5. Mary Lou Williamson, ed., *Greenbelt: History of a New Town, 1937–1987* (Norfolk, Va.: Donning, 1987), 117; Arnold, *The New Deal in the Suburbs*, 221–22. When the federal government sold the defense housing and it became part of a housing cooperative, the management added landscaping as well as some interior sidewalks.

6. Weekly issues of the *Cooperator* described each stage of construction and the moving in of new families.

7. Arnold, *The New Deal in the Suburbs*, 220.

8. The *Cooperator* provided details of each community holiday event as well as information on special events for the new residents.

9. *Cooperator*, February 5, 1943.

10. Weekly issues of the *Cooperator* discussed each development and change in the co-op businesses.

11. Williamson, *Greenbelt*, 117–21.

12. *Washington Post*, October 6, 1944.

13. *Cooperator*, April 20, 1945.

14. *Cooperator*, May 8, 1942.

15. Folder T, Box 2961, Eleanor Roosevelt Papers, Franklin Roosevelt Presidential Library, Hyde Park, N.Y.

16. Data for Greenbelt from the *Cooperator*. National data from U.S. Bureau of the Census, *Statistical Abstract of the United States*, 93d ed. (Washington, D.C., 1972), 373.

17. *Cooperator*, August 1, 1940.

18. Patricia Dunbar Gray, Edith Frank, and Ruth Bowman, in *Looking Back: Greenbelt Is 50, 1937–1987* (Greenbelt, Md.: City of Greenbelt, 1987), 18, 26, 39.

19. Arnold, *The New Deal in the Suburbs*, 222–23; Williamson, *Greenbelt*, 118–22.

20. Williamson, *Greenbelt*, 118–20.

21. Philip S. Brown, "What Has Happened at Greenbelt?" *New Republic,* August 11, 1941, 183–85; Ed Thornhill and Fred De Armond, "Another Social Experiment Goes Sour: Greenbelt for Sale," *Nation's Business* 28 (October 1940), 23; "Everyone Owns All Businesses in Co-op City," *Christian Science Monitor,* January 29, 1940; "Residents Acquire Greenbelt Co-op," *New York Times,* January 28, 1940.

22. *Washington Post,* July 21 and September 6, 1940; July 3, 1941; October 2, 1944.

23. *Washington Daily News,* August 13, 1940, March 16, 1942; *Washington Post,* March 12, 1942; *Washington Star,* March 12, 1942.

24. *Washington Post, Washington Star,* and *Washington Times-Herald,* May 29, 1943. The story was picked up by the Associated Press wire service and circulated to papers throughout the country.

25. Mrs. Wells Harrington, Anthony Madden, and Lolita Granahan, in *Looking Back,* 99, 45, 34.

26. Mary Clare Bonham England, in *Looking Back,* 27.

27. *Cooperator,* August 17, 1945.

28. *Cooperator,* October 6, 1944.

## 4 / The Government versus Greenbelt

1. Quoted in Page Smith, *As a City Upon a Hill: The Town in American History* (New York: Knopf, 1968), 8.

2. Bert Wissman, "House Begins Night Sessions in Probe of Greenbelt Co-ops," *Washington Post,* August 23, 1947.

3. According to the Legislative Branch of the National Archives, none of the records of this investigation are in its holdings; the only sources available are the *Cooperator* and the Washington newspapers.

4. The *Cooperator* discussed rumors of the sale in articles at the end of 1945, forcing the city council to reveal the secret negotiations.

5. Paul Conkin, *Tomorrow a New World: The New Deal Community Program* (Ithaca: Cornell University Press, 1959), 229.

6. Joseph Arnold, *The New Deal in the Suburbs: A History of the Greenbelt Town Program, 1935–1954* (Columbus: Ohio State University Press, 1971), 230–32.

7. K. C. Parsons, "Clarence Stein and the Greenbelt Towns," *APA Journal,* vol. 56 (Spring 1990): 161–62.

8. Ibid., 178.

9. Arnold, *The New Deal in the Suburbs,* 234–35.

10. *Looking Back: Greenbelt Is 50, 1937–1987* (Greenbelt, Md.: City of Greenbelt, 1987); several individuals who discussed this time period with the author recalled the prominence of pacifists in town after the war.

11. *Cooperator,* September 14, 1945.

12. James Giese, city manager, interview by author, May 19, 1990.

13. Henry Klumb's plans can be found in the Tugwell Room, Greenbelt Public Library.

14. *Cooperator,* November 9, 1945.

15. *Cooperator,* July 19, 1946.

16. Various of Hale Walker's designs for Greenbelt's completion, plus Oliver Winston's and Clarence Stein's ideas for the future development of the town put forth at the time of its sale, exist in National Archives II, College Park, Maryland.

17. F. H. C. Company, "General Plan of Land Use East of Edmonston-Beltsville Road" (1955); only Sheet A, the upper half, remains (Tugwell Room).

18. Conkin, *Tomorrow a New World*, 324–25; information from Dorothy Lauber, secretary to the city manager of Greenbelt. The total cost had been $14,016,270, not including the defense homes. However, the original cost had been grossly inflated by the use of relief labor, as one of the primary aims had been to provide jobs for the unemployed. See Arnold, *The New Deal in the Suburbs*, 195.

19. Weekly issues of the *Cooperator* throughout this period described the sale, the loan from the Farm Bureau Insurance Companies, and all other aspects of the financial maneuvering necessary for the GVHC to complete the purchase.

20. Early GVHC activist who still lives in Greenbelt, interview by author, May 26, 1989.

21. "Prospective Buyers Flock to Greenbelt," *Washington Post*, November 10, 1952.

22. An examination of the real estate sections of the Sunday issues of the *Washington Post–Times-Herald* for two weeks before the Greenbelt advertisement (May 1 and 8, 1955) showed prices of new "moderate size" homes in Kensington for $26,500, while "moderate size" new ramblers sold for $21,950 in Arlington, Va. A new development in Rockville, advertised as "the best deal on the market today," had homes for $14,990. Prices in Prince George's County were somewhat lower, with new homes advertised for $12,300. The only prices as low as those in Greenbelt were homes in Manassas Park, Va. a good distance from Washington, D.C., for $7,000. However, it must be remembered that the units in Greenbelt were small row houses, which thus should have cost less than even modest freestanding homes. No row houses were advertised to make a direct comparison to Greenbelt possible.

23. *Cooperator*, April 30, 1953.

24. Ibid.; *Cooperator*, May 21, 1953.

25. *Cooperator*, June 21 and December 20, 1951.

26. Mary Lou Williamson, ed., *Greenbelt: History of a New Town, 1937–1987* (Norfolk: Donning, 1987), 157.

27. Efforts began with the Dies Committee in 1938, followed by the Foreign Agents Registration Act, also in 1938. The 1939 Hatch Act denied federal jobs to members of any organization or party that advocated overthrowing the government. The 1940 Voorhis Act demanded that all groups with foreign affiliations register with the government. During World War II, the secretaries of war and the navy could dismiss summarily any worker whose continued employment was deemed contrary to national security. Richard M. Fried, *Nightmare in Red: The McCarthy Era in Perspective* (New York: Oxford University Press, 1990).

28 Fried, *Nightmare in Red*, 66–67.

29. George H. Callcott, *Maryland and America, 1940–1980* (Baltimore: Johns Hopkins University Press, 1985), chap. 5.

30. Anthony Lewis, "You Get the Feeling People Are Crawling Out of the Woodwork," *Washington Daily News*, April 19, 1954.

31. Anthony Lewis, "Many Things Hinge on Navy's Case against Greenbelt 'Risk,'" *Washington Daily News,* April 16, 1954; Williamson, *Greenbelt,* 154.

32. Bruce Bowman, in *Looking Back,* 31.

33. Anthony Lewis, "Thank God It Happened to You," *Washington Daily News,* April 21, 1954. Both the National Archives and the Office of the Judge Advocate General of the Department of the Navy disclaim knowledge of any records resulting from this investigation, so factors such as the identity of Chasanow's accusers remain unknown.

34. Lewis, "Thank God It Happened to You."

35. Anthony Lewis, "Find Compassion in Your Heart," *Washington Daily News,* April 22, 1954.

36. Bruce Bowman, in *Looking Back,* 31.

37. Joseph P. Blank, *Look,* May 17, 1955, 28.

38. Ibid., 26.

39. Lewis, "You Get the Feeling."

40. Anthony Lewis, "Rank Amateurs in the Field of Security Investigations," *Washington Daily News,* April 20, 1954.

41. Lewis, "Thank God It Happened to You."

42. Lewis, "Many Things Hinge on Navy's Case."

43. Blank, *Look,* 28.

44. "Services Terminated," *Washington Star,* April 17, 1954.

45. "Navy to Give Chasanow Chance to Present Appeal Case," *Washington Post,* April 19, 1954.

46. "Chasanow Informed 2 Charges Dropped," *Washington Post,* June 5, 1954; "Chasanow Cleared and Restored to Job; Navy Admits Error," *Washington Star,* September 1, 1954.

47. Anthony Lewis, "On the Occasion of Your Retirement," *Washington Daily News,* July 27, 1955.

48. Williamson, *Greenbelt,* 151; individual closely involved with the case who wishes to remain anonymous, interview by author, March 8, 1990.

49. *Greenbelt News Review,* October 25, 1956.

50. Early resident, who was a leader of the GVHC, interview by author, May 26, 1989.

51. *Cooperator,* January 13, 1949.

52. *News Review,* December 9, 1954.

53. "Pastor Building Greenbelt Church Faces 'Pioneer' Problems," *Washington Post,* November 2, 1946.

54. Caspar Nannes, "Catholics and Protestants Help Build Greenbelt Jewish Center," *Washington Star,* May 24, 1953.

55. Frieda Perlzweig, Eunice Birtman Burton, and Ben Rosenzweig, in *Looking Back,* 67, 59, 60, 68.

56. The February and March 1952 issues of the *Cooperator* printed details on the formation of the co-op kindergarten.

57. "Greenbelt OK's Sale of School," *Washington Post,* July 15, 1958.

58. School details from the *Cooperator.*

59. *Washington Post,* June 8, 1947.

60. Sam Zagoria, "Greenbelt Children Featured as Town Marks 10th Birthday," *Washington Post*, August 8, 1947.

61. Remarks at Greendale's Decennial Celebration, Box 63, Tugwell Collection, Franklin Roosevelt Presidential Library, Hyde Park, N.Y.

62. Lowell Mellett, "Times Takes the Laugh Out of That Funny Word Tugwelltown," *Washington Star*, August 5, 1947; "Greenbelt: Where the Library Collects More Fines Than the Courts," July 28, 1947; *Washington Post*, August 8, 1947.

63. Carol Willis Moffett, "They Heard the Hungry," *Toronto Star Weekly*, August 3, 1946.

64. No data on local or city elections could be found to compare with Greenbelt's percentage of eligible voters who actually voted, so it cannot be determined if Greenbelt's numbers are typical. Apparently the only way to find such data is through local newspapers, such as this from the *Cooperator* and later the *News Review*.

65. *Cooperator*, September 21, 1945; U.S. Bureau of the Census, figures in Michael Fischetti Jr. "The Development of the City of Greenbelt, Maryland: Its Growth, Government, and Politics" (Master's thesis, University of Maryland, 1967), 100; *Cooperator*, October 6, 1949.

66. *News Review*, November 8, 1956.

67. Material on the referendum from the *News Review*, which followed changes in local government structure.

68. "Teen-Agers Lobby to Get Skating Back," *Washington Post*, September 10, 1958.

69 Material on rent increases from issues of the *Cooperator* of the period.

70. *Cooperator*, January 21, 1954.

71. *Cooperator*, May 6 and September 30, 1954.

72. "Tugwell Planning Home in Greenbelt He Created," *Washington Star*, October 1955; *News Review*, October 13, 1955.

73. Zane L. Miller, *Suburb, Neighborhood, and Community in Forest Park, Ohio, 1935–1976* (Knoxville: University of Tennessee Press, 1981), 3.

## 5 / Developers versus Greenbelt

1. Material on the growth of Greenbelt from the *News Review*, which covered the town's physical changes in detail.

2. "Greenbelt Park Play Area to Start in July," *Washington Post*, June 1, 1958.

3. *News Review*, August 7, 1958.

4. *News Review*, December 27, 1956.

5. "Greenbelt Gets Guide to Future," *Washington Post–Times-Herald*, January 17, 1957.

6. "Residents Rap Greenbelt Plan," *Washington Star*, February 28, 1957.

7. Ibid.

8. *News Review*, April 19, 1962; Citizens for a Planned Greenbelt File, Tugwell Room, Greenbelt Public Library.

9. *News Review*, July 5, 1962.

10. *News Review*, July 7, 1962.

11. Material on development is from the *News Review* unless otherwise specified.

12. *News Review*, July 13, 1961.

13. *News Review*, September 19, 1963.

14. Material on development is from the *News Review*. In this period many Greenbelters jokingly referred to the front page of the paper as the "Bresler page," due to the prominence given to development issues every week.

15. Michael Fischetti Jr. "The Development of the City of Greenbelt, Maryland: Its Growth, Government, and Politics" (Master's thesis, University of Maryland, 1967), 63.

16. Peter Masley, "Master Plan for Greenbelt Assailed by Citizen Groups," *Washington Star*, July 22, 1964.

17. "Greenbelt Creates Own Master Plan, Adopts It," *Washington Post*, March 8, 1965.

18. *News Review*, March 18, 1965.

19. Citizens for a Planned Greenbelt File, Tugwell Room, Greenbelt Public Library.

20. Fischetti, "The Development of the City of Greenbelt," 32.

21. *News Review*, October 21, 1965.

22. Fischetti, "The Development of the City of Greenbelt," 77–81.

23. Ibid., 82.

24. *News Review*, April 7, 1966.

25. *News Review*, April 21, 1966.

26. Citizens for a Planned Greenbelt File, Tugwell Room.

27. *News Review*, July 28, 1966.

28. Leonard Downie Jr., *Mortgage on America* (New York: Praeger, 1974), 119.

29. Ibid., 121.

30. *News Review*, August 18, 1966.

31. *News Review*, September 15, 1966.

32. *News Review*, December 7, 1967.

33. *News Review*, September 29, 1966.

34. *News Review*, August 12, 1966.

35. *News Review*, January 19, 1967.

36. *News Review*, June 8, 1967.

37. *News Review*, April 6, 1967.

38. *News Review*, October 26, 1967.

39. *News Review*, September 21, 1967.

40. Michael Drosnin, "Bresler Libel Suit against Greenbelt Newspaper Opens," *Washington Post*, January 4, 1968.

41. *News Review*, January 11, 1968.

42. Drosnin, "Bresler Libel Suit against Greenbelt."

43. *News Review*, January 11, 1968.

44. Ibid.

45. *News Review*, January 18, 1968.

46. "Greenbelt Citizens Urge Paper to Appeal," *Washington Post*, January 27, 1968.

47. *News Review*, February 1, 1968.

48. *News Review*, February 22, 1968.

49. The Abraham Chasanow and Isidore Parker who were now supporting opposite sides in the *News Review* case are, indeed, the same men once deemed security risks by the navy and fired from their jobs.

50. *News Review*, April 11, 1968.

51. *News Review,* August 15, August 22, and September 12, 1968.

52. *News Review,* December 26, 1968.

53. *News Review,* December 19, 1968.

54. U.S. Bureau of the Census; Mary Lou Williamson, ed., *Greenbelt: History of a New Town, 1937–1987* (Norfolk, Va.: Donning, 1987), 196.

55. George H. Callcott, *Maryland and America, 1940–1980* (Baltimore: Johns Hopkins University Press, 1985), 65.

56. Downie, *Mortgage on America,* 119–32.

57. *News Review,* May 8, 1969. Anthony Lewis, *Make No Law: The Sullivan Case and the First Amendment* (New York: Vintage, 1992), examines the *New York Times* case.

58. *News Review,* May 8, 1969.

59. Subcommittee on Oral History, Greenbelt Fiftieth Anniversary Committee; videotape in private collection.

60. Supreme Court of the United States, no. 413, October 1969 term, Greenbelt Cooperative Publishing Association, Inc. et al., Petitioners, v. Charles S. Bresler.

61. *News Review,* July 16, 1970.

62. *New York Times,* May 19, 1970; *News Review,* June 18, 1970; John P. MacKenzie, "Libel Case Award Is Nullified," *Washington Post,* May 19, 1970.

63. *Washington Evening Star,* May 18, 1970; Associated Press release, May 19, 1970, printed in *News Review,* November 30, 1970.

64. Dinner speeches, reported in *News Review,* December 17, 1987.

65. *News Review,* March 12, 1970.

66. *Washington Evening Star,* January 4, 1976.

67. *News Review,* March 14, 1968.

68. Address by Al Herling, president of Friends of the Greenbelt Library, given at the library dedication on April 7, 1979. Friends of the Greenbelt Library File, Tugwell Room.

69. Program of dedication ceremony, Woman's Club File, Tugwell Room.

70. Ben H. Bagdikian, "The Rape of the Land," *Saturday Evening Post,* June 18, 1966, 27; Ken Schlossberg, "Greenbelt Has Been Deflowered," *Washingtonian Magazine* (July 1968): 49–51, 75–76.

71. Data from U.S. Bureau of the Census.

72. *The Neighborhoods of Prince George's County,* Final Report, Project MDR-54 (CR)(Prince George's County: December 1974), 302.

73. Charles Krause, "Sewer Intercession Denied by Mandel," *Washington Post,* September 23, 1973; *News Review,* September 27, 1973.

74. Material is from weekly issues of the *News Review.*

75. Ebenezer Howard, *Garden Cities of Tomorrow,* edited by F. J. Osborne (Cambridge: MIT Press, 1965), suggested that his model town both house workers and provide them workplaces. A number of British towns were built using his model. Some scholars maintain that greenbelt towns originated from the garden city concept.

76. *Looking Back: Greenbelt Is 50, 1937–1987* (Greenbelt, Md.: City of Greenbelt, 1987), 113.

77. *News Review,* June 7, 1982.

78. Elizabeth Rathbun, "Greenbelt: When the Old Meets the New," *Prince George's Journal,* August 16, 1983.

79. *News Review,* August 9, 1984.

80. James Giese, city manager, interview by author, May 19, 1990.

81. *News Review,* November 7, 1985.

## 6 / Overcoming Difficulties in Cooperation

1. Material on GCS is from weekly issues of the *News Review.* Readers criticized the newspaper for placing too much emphasis on GCS affairs.

2. *News Review,* June 21 and July 5, 1956.

3. Phil Jacobs, "Co-op Survives in Area Market," *Prince George's Journal,* August 9, 1978.

4. *News Review,* January 5, 1984.

5. *News Review,* February 9, 1984.

6. Jim Cassels, Oral History Subcommittee, Greenbelt Fiftieth Anniversary Committee; videotape in private collection.

7. The turmoil among Greenbelt Homes members caused by the need for rehabilitation received detailed coverage every week in the *News Review.*

8. Michael Knepler, "Paid Up!" *Washington Star,* March 1, 1978.

9. *News Review,* February 2, 1978.

10. *News Review,* September 28, 1967.

11. Kaye Sizer Noe, "The Fair Housing Movement; An Overview and a Case Study" (master's thesis, University of Maryland, 1965), 149.

12. Bruce Bowman, in *Looking Back,* 36–37.

13. In 1970 Greenbelt's black homeowners lived in houses with an average value of $35,019, while white homeowners' average home value was only $27,550. Black renters paid an average of $200 per month; white renters paid an average of $162. *The Neighborhoods of Prince George's County,* Final Report, Project MDR-54 (CR)(Prince George's County: December 1974).

14. The summary of the problems of early integration efforts in Greenbelt is based on information from several individuals active in Greenbelt Citizens for Fair Housing, interviews by author.

15. Noe, "The Fair Housing Movement," 149.

16. *News Review,* January 4, 1973.

17. *News Review,* January 25 and January 18, 1973.

18 George H. Callcott, *Maryland and America, 1940–1980* (Baltimore: Johns Hopkins University Press, 1985), 246.

19. Springhill Lake and Beltway Plaza consistently had the highest crime rates in Greenbelt. In 1978 the total of all crimes reported to police was 300 in Springhill Lake and Beltway Plaza, 198 in central Greenbelt, and 82 in Greenbelt East. *News Review,* April 26, 1979. In 1990 Springhill Lake and Beltway Plaza reported 699 crimes, Greenbelt East 471, and central Greenbelt 331. Greenbelt Police Department.

20. Material from *News Review.*

7 / The Persistence of the Greenbelt Idea

1. Outstanding Citizen Award File, Tugwell Room, Greenbelt Public Library.

2. Labor Day Festival File, Tugwell Room.

3. *News Review,* June 7, 1979.

4. *News Review,* November 29, 1973.

5. Mary Lou Williamson, ed., *Greenbelt: History of a New Town, 1937–1987* (Norfolk, Va.: Donning, 1987), 225.

6. Resident who moved to Greenbelt in 1985, interview by author, March 10, 1990.

7. *News Review,* November 1976 issues.

8. Marjorie Hyer, "Greenbelt Marks Religious Holidays with Interfaith Walk," *Washington Post,* April 18, 1987.

9. *News Review,* December 1, 1977.

10. *News Review,* January 29, 1970.

11. Williamson, *Greenbelt,* 226.

12. Deborah Dash Moore described Caroline Ware's unique approach to research in her introduction to Ware's *Greenwich Village, 1920–1930: A Comment on American Civilization in the Post-War Years* (Berkeley: University of California Press, 1944). Ware combined a cultural approach to history with the methods of anthropology, producing a vivid analysis of change in a multiethnic neighborhood. Ware was concerned with the future of the city and saw the importance of finding common values among groups, particularly in the American culture, which emphasizes the individual.

13. Resolution 550, passed by the city council on July 11, 1983, effective July 21, 1983. Greenbelt Fiftieth Anniversary Committee File, Tugwell Room.

14. Members of subcommittees of Greenbelt Fiftieth Anniversary Committee, interviews by author.

15. Minutes of committee meetings, Greenbelt Fiftieth Anniversary Committee File, Tugwell Room.

16. Greenbelt Museum Committee File, Tugwell Room.

17. Mary Lou Williamson to Betty Allen, Greenbelt Fiftieth Anniversary Committee File, Tugwell Room,

18. New Towns Conference File, Tugwell Room.

19. *News Review,* August 6 and May 28, 1987.

20. Larry Van Dyne, "How the New Deal Changed Washington," *Washingtonian* (February 1982): 122; Michael Eastman, "A 'Green Town' Built in Thirties Battles to Stay Model Community," *Washington Post,* September 20, 1979; Sara Rimer, "Greenbelt: U.S. Dream Town," *Washington Post,* June 30, 1981. Quotation from Eastman article.

21. Peter Ruehl, "Greenbelt: 'Ruined City' or 'Still the Best Place to Live'?" *Prince George's Sentinel,* April 8, 1971; Peter D. Pichaske, "Model City Enters the 1980's," *Prince George's Journal,* August 18, 1982.

22. David Margolick, "A Suburb Recalls Its New Deal Mission," *New York Times,* June 9, 1985; *News Review,* November 29, 1984.

23. *Congressional Record,* July 14, 1987.

24. Lloyd W. Bookout, "Greenbelt, Maryland: A New Town Turns 50," *Urban Land*

(August 1987), 10–12. Also see Laurence E. Coffin Jr. and Beatriz de Winthuysen Coffin, "Greenbelt: A Maryland New Town Turns 50," *Landscape Architecture* (June 1988), 48–53.

25. *News Review,* May 28, 1987.

26. *News Review,* December 28, 1978.

27. *News Review,* May 15, 1986.

28. *News Review,* May 30, 1985.

## 8 / Greenbelt Today and Tomorrow

1. Remarks for Outstanding Citizen Presentation, 1988, Outstanding Citizen File, Tugwell Room, Greenbelt Public Library.

2. Michele L. Norris, "School Was Only Haven of Hope," *Washington Post,* July 31, 1989.

3. "Dooney's Story," *Washington Post,* August 3, 1989.

4. Virtually all Greenbelt residents interviewed by author expressed the view that the tone of the paper changed in reaction to the Supreme Court suit. They felt that the paper carried fewer editorials than before and was more cautious. The only residents disagreeing with this view were newspaper staff.

5. *News Review,* November 21, 1996.

6. Susan Gervasi, "Greenbelt to Get U.S. Aid," *Washington Post,* December 14, 1989.

7. *News Review,* December 17, 1987.

8. Susan Gervasi, "East-West Tension Builds in Greenbelt," *Washington Post,* April 7, 1988.

9. Susan Gervasi, "An Issue of Living Space," *Washington Post,* November 2, 1989.

10. *News Review,* June 21, 1990.

11. Susan Gervasi, "Wetlands Plan Called Harmful," *Washington Post,* January 11, 1990.

12. Wendy Melillo, "Greenbelt Residents Jump to Defend Its Precious Belt," *Washington Post,* September 28, 1996.

13. Howard S. Berger and Robert D. Rivers, "Greenbelt Historic District Study" (Maryland–National Capital Park and Planning Commission, 1994).

14. *News Review,* June 6, 1996.

15. *News Review,* May 20, 1999; the city council had not pursued the issue as of November 1999.

16. Mary Lou Williamson, ed., *Greenbelt: History of a New Town, 1937–1987* (Norfolk, Va.: Donning, 1987), 220.

17. *News Review,* June 29, 1989.

18. *News Review,* March 21, 1996.

19. Ibid.

20. *News Review,* March 14, 1996.

21. Williamson, *Greenbelt,* 312; *News Review,* March 6, 1997.

22. John C. Buckner, "The Development of an Instrument and Procedure to Assess the Cohesiveness of Neighborhoods" (Ph.D. diss, University of Maryland, 1986).

23. James Giese, city manager, interview by author, May 19, 1990.

24. *News Review,* February 24, 1966.

25. This perception is based on conversations with residents of Old Greenbelt.

26. *News Review,* June 10, 1982.

27. *News Review,* December 1, 1977.

28. *News Review,* May 24, 1979.

29. *News Review,* March 13, 1986.

30. Described in several issues of *News Review,* May 1982.

31. *News Review,* July 5, 1984.

32. *News Review,* January 23, 1986.

33. *News Review,* January 26, 1989.

34. *News Review,* March 31, 1988.

35. *News Review,* December 27, 1990.

36. Giese, interview by author.

37. "A New Year of Green Line Construction," *Washington Post,* January 3, 1991; "Lukewarm Thrill at End of Line, New Metro Station Fails to Dazzle All in Greenbelt," *Washington Post,* December 10, 1993.

38. Brochure in honor of Elaine Skolnik, produced for the fiftieth anniversary celebration of the *News Review.*

39. Longtime Greenbelt resident, interview by author, March 8, 1990.

40. *News Review,* October 11, 1973; Vivian Rigdon, "Greenbelt, Despite Changes, Retains Cooperative Spirit," *Prince George's Sentinel,* July 28, 1976; Michael Knepler, "Paid Up! Another Spur to Greenbelt Pride," *Washington Star,* March 1, 1978.

41. Peter D. Pichaske, "Model City Enters the 1980s," *Prince George's Journal,* August 18, 1982.

42. Ibid.

43. Joan McQueeny Mitric, "Hundreds Come Home to Greenbelt," *Washington Post,* October 6, 1982.

44. "Cooperative Spirit Lives on in Greenbelt," *Prince George's Journal,* October 7, 1987.

45. Resident of Greenbelt Homes who moved to Greenbelt in 1985, interview by author, March 20, 1990.

46. Resident who moved into Lakeside North in 1989, interview by author, January 5, 1991.

47. These bulletins can be found in Record Group 287, National Archives, Washington, D.C., and in some federal depository libraries.

48. The Parkwood subdivision, home of the author and divided by the wisdom of the U.S. Postal Service into Bethesda and Kensington, Maryland, outside of Washington, D.C., provides a good example. One of the earliest postwar subdivisions in the area, Parkwood has local, curving streets surrounded by larger roads and Rock Creek Park. An elementary school with a park is somewhat centrally located in the neighborhood.

49. Philip Langdon, *A Better Place to Live: Reshaping the American Suburb* (Amherst: University of Massachusetts Press, 1994), xii–xiii.

50. Peter Katz, *The New Urbanism: Toward an Architecture of Community* (New York: McGraw-Hill, 1994), xvii.

51. William Fulton, *The New Urbanism: Hope or Hype for American Communities?* (Cambridge, Mass.: Lincoln Institute of Land Policy, 1996).

52. Lewis Mumford, *The Culture of Cities* (Westport, Conn.: Greenwood Press, 1970 [1938]).

# Bibliographical Essay

As there is no previous scholarly work on Greenbelt, I have relied largely on primary sources to write its story. The foremost of these is the town newspaper, published weekly, which serves as a diary of the community. Called the *Cooperator* in its first years, it is now the *Greenbelt News Review*. Other area newspapers, in particular the *Washington Post* and the *Washington Evening Star*, frequently focused on events in town, reflecting the politics of the day in their coverage.

Manuscript collections served as a major resource for this work, especially the one formed by Greenbelters themselves, now in the Tugwell Room of the Greenbelt Public Library. The Tugwell Room contains papers, souvenirs, and recollections of the first Greenbelters. The Franklin Roosevelt Presidential Library, in Hyde Park, New York, maintains numerous collections pertinent to Greenbelt's formative years, among them the Eleanor Roosevelt Papers, the Tugwell Collection, the Morganthau Papers, and the Presidential Secretaries File. The University of Kentucky Library in Lexington, Kentucky, houses the Lansill Manuscript Collection, papers donated by John S. Lansill, the head of the Suburban Division of the Resettlement Administration. Studying Greenbelt is somewhat of a cottage industry at the nearby University of Maryland, and students have produced several theses and dissertations on the subject. Particularly useful is William Form's 1944 study, "The Sociology of a White Collar Suburb: Greenbelt, Maryland."

Greenbelt began its life as a project of the federal government, so federal records on its creation exist in abundance. The National Archives II, in College Park, Maryland, contains a variety of material on the planning of the town as well as detailed blueprints and paperwork on every aspect of its construction. These are contained in Record Groups 16, 96, 121, and in eight portions of 196, and 207. The famed photographers of the Farm Security Administration visited Greenbelt at several points during its construction and in its first years. Their photographs can be found in the Farm Security Administration Collection, Prints and Photographs Division, Library of Congress, Washington, D.C.

The views of Greenbelt residents form a necessary part of any story about their city. I therefore interviewed a number of them for this work. Previous interviews include those by University of Maryland students in 1981–85, published in *Looking Back: Greenbelt Is 50, 1937–1987* (1987). Videotapes, which are privately owned, were made as part of Greenbelt's fiftieth anniversary celebration.

The major previous work on Greenbelt, along with its sister cities, Greenhills, Ohio, and Greendale, Wisconsin, is Joseph Arnold's *The New Deal in the Suburbs: A History of the Greenbelt Town Program, 1935–1954* (1971). Additional works on Greenbelt are Otis Kline Fulmer, *Greenbelt* (1942), a bibliography by Susan L. Klaus, *Links in the Chain:*

*Greenbelt, Maryland, and the New Town Movement in America* (1987), George Warner, *Greenbelt, the Cooperative Community: An Experience in Democratic Living* (1954), Christian Larsen and Richard Andrew, *The Government of Greenbelt* (1951), as well as the "picture book" produced by the city in honor of its fiftieth anniversary celebration: *Greenbelt: History of a New Town, 1937–1987,* edited by Mary Lou Williamson (1987). Arnold Alanen and Joseph Eden have written on Greendale in *Main Street Ready-Made: The New Deal Community of Greendale, Wisconsin* (1987); Zane Miller's *Suburb: Neighborhood and Community in Forest Park, Ohio, 1935–1976* (1981) describes the early years of Greenhills, focusing mainly on Forest Park.

Most valuable for putting Greenbelt in its immediate context was Clarence Stein, *Toward New Towns for America* (1966). Also useful were Paul Conkin, *Tomorrow a New World: The New Deal Community Program* (1959), Walter Creese, *The Search for Environment: The Garden City, Before and After* (1966), James Dahir, *The Neighborhood Unit Plan, Its Spread and Acceptance: A Selected Bibliography with Interpretive Comments* (1947), and of course Ebenezer Howard, *Garden Cities of Tomorrow* (1965). While these works helped with the early twentieth century, John Reps provided the overall context in his *The Making of Urban America: A History of City Planning in the United States* (1965). Also useful in this vein is Daniel Schaffer, *Two Centuries of American Planning* (1988).

To consider the origins of Greenbelt as a response to the crisis of the Great Depression, and to adequately understand what it meant for those who became its residents, I turned to works on the Depression, the New Deal, and the experiences of those who lived through both. Useful for understanding the long-term impact of the New Deal was Steve Fraser and Gary Gerstle's *The Rise and Fall of the New Deal Order, 1930–1980* (1989). Also helpful were Charles Beard and George Smith, *The Future Comes: A Study of the New Deal* (1972), Phoebe Cutler, *The Public Landscape of the New Deal* (1985), Robert Goldston, *The Great Depression: The United States in the Thirties* (1968), and Katie Louchheim, *The Making of the New Deal* (1983). James Patterson's *Congressional Conservatism and the New Deal* (1967) describes the difficulties experienced by the Resettlement Administration.

Several oral histories provided the first-person experience of the time period, especially Edward Anderson, *Hungry Men* (1935), but also Studs Terkel, *Hard Times: An Oral History of the Great Depression* (1970), and Tom Terrell and Jerrold Hirsch, *Such as Us: Southern Voices of the Thirties* (1978). Works such as Ronald Grele, *Envelopes of Sound: The Art of Oral History* (1985), and Laurence C. Watson and Maria-Barbara Watson-Franke, *Interpreting Life Histories: An Anthropological Inquiry* (1985), guided me in my oral history approach.

The most useful work I found on cooperatives was Joseph G. Knapps' two-volume set: *The Rise of American Cooperative Enterprise, 1620–1920* and *The Advance of American Cooperative Enterprise, 1920–1945* (1973). The research of sociologists provided useful background for analyzing Greenbelt: Terry Nichols Clark, *Community Power and Policy Outputs: A Review of Urban Research* (1973), James S. Coleman, *Community Conflict* (1957), Robert A. Dahl, *Who Governs? Democracy and Power in an American City* (1961), Claude S. Fischer, *To Dwell among Friends: Personal Networks in Town and City* (1982), Robert Fisher, *Let the People Decide: Neighborhood Organizing in America* (1984), Floyd

Hunter, *Community Power Structure: A Study of Decision Makers* (1953), C. Wright Mills, *The Power Elite* (1956), Thomas Bender, *Community and Social Change in America* (1982), and Nelson W. Polsby, *Community Power and Political Theory* (1963).

The largest body of material necessary to place Greenbelt into its proper context was the literature on planning, which can be divided into several types. Because early detractors frequently derided Greenbelt as utopian, I explored the concept of utopia. The goal of creating an ideal environment probably began very early in human history and is reflected well in Thomas More's *Utopia* (edited by Edward Surtz, 1964). Americans, especially in the nineteenth century, attempted to establish their own utopian communities, as explained in *Utopias: The American Experience*, edited by Gairdner B. Moment and Otto F. Kravshaar (1980). John Garner explored another type of search for an ideal environment in *The Model Company Town: Urban Design through Private Enterprise in Nineteenth-Century New England* (1984). Works such as Robert Fishman, *Urban Utopias in the Twentieth Century: Ebenezer Howard, Frank Lloyd Wright, and Le Corbusier* (1977), follow the concept into our times.

Early in the twentieth century Ebenezer Howard's garden city caught the fancy of planners in England and the United States, as explicated in Stanley Buder, *Visionaries and Planners: The Garden City Movement and the Modern Community* (1990). British planner Frederic J. Osborn discussed garden cities in *Greenbelt Cities* (1969), while Carol Christensen covered both garden cities and their newer reincarnation in *The American Garden City and the New Towns Movement* (1986). The new towns of the 1960s received much attention in works such as James Bailey, *New Towns in America: The Design and Development Process* (1971), Gideon Golany and Daniel Walden, editors, *The Contemporary New Communities Movement in the United States* (1974), Raymond J. Burby and Shirley F. Weiss, *New Communities, U.S.A.* (1976), Carol Corden, *Planned Cities: New Towns in Britain and America* (1977), Donald C. Klein, editor, *Psychology of the Planned Community: The New Town Experience* (1978), and in Britain, Lloyd Rodwin, *The British New Towns Policy: Problems and Implications* (1956).

As suburbs became an increasingly important part of American life, planners and historians turned to their study. Some produced general works, such as Robert Fishman's *Bourgeois Utopias: The Rise and Fall of Suburbia* (1987), and Robert A. M. Stern's *The Anglo-American Suburb* (1981), while many focused on the story of a particular suburb, such as Michael H. Ebner, *Creating Chicago's North Shore: A Suburban History* (1988), and Carol A. O'Connor, *A Sort of Utopia: Scarsdale, 1891–1981* (1983). John Stilgoe focused on the history of suburbs in *Borderland: Origins of the American Suburb, 1820–1939* (1988).

At the end of the twentieth century some planners turned their attention away from the ever-expanding suburbs to focus on the faltering core of cities, as the new urbanism took hold. Representative works on the topic are Peter Katz, *The New Urbanism: Toward an Architecture of Community* (1994), Philip Langdon, *A Better Place to Live: Reshaping the American Suburb* (1994), and William Fulton, *The New Urbanism: Hope or Hype for American Communities?* (1996). The irony of the new urbanism is that its ideas are based on old-fashioned, traditional American towns of the type frequently demolished during the 1960s' urban renewal.

The study of a community like Greenbelt thus requires an understanding of an unusual variety of topics, from the history of planning to sociological studies to works on cooperatives. This range drew me to the study of Greenbelt in the first place, reflecting as it does the complexity of real life and real people in a community planned to be as perfect as possible.

# Index

# About the Author

Cathy Knepper was born in Wichita, Kansas, and grew up in Philadelphia, Denver, and points in-between. She received her BA in English from the University of Michigan, her master's in social work at Case Western Reserve University, and her doctorate in American studies from the University of Maryland at College Park. She is an independent scholar residing in Kensington, Maryland, and works for Amnesty International.

# Titles of Related Interest in the Series